JOINT SOVIET-AMERICAN WORKSHOP ON THE PHYSICS OF SEMICONDUCTOR LASERS

AIP CONFERENCE PROCEEDINGS 240

JOINT SOVIET-AMERICAN WORKSHOP ON THE PHYSICS OF SEMICONDUCTOR LASERS

LENINGRAD, USSR 1991

EDITOR:
ZHORES I. ALFEROV
A. F. IOFFE PHYSICO-TECHNICAL INSTITUTE

American Institute of Physics　　　　　　　　　　**New York**

A. F. Ioffe Physico-Technical Institute, Leningrad, USSR JV "SMART", Leningrad, USSR

This book was composed using T_EX. The editing and CRC were produced by the Editorial office at the Ioffe Physico-Technical Institute.

Authorization to photocopy items for internal or personal use, beyond the free copying permitted under the 1978 U.S. Copyright Law (see statement below), is granted by the American Institute of Physics for users registered with the Copyright Clearance Center (CCC) Transactional Reporting Service, provided that the base fee of $2.00 per copy is paid directly to CCC, 27 Congress St., Salem, MA 01970. For those organizations that have been granted a photocopy license by CCC, a separate system of payment has been arranged. The fee code for users of the Transactional Reporting Service is: 0094-243X/87 $2.00.

© 1992 American Institute of Physics.

Individual readers of this volume and nonprofit libraries, acting for them, are permitted to make fair use of the material in it, such as copying an article for use in teaching or research. Permission is granted to quote from this volume in scientific work with the customary acknowledgment of the source. To reprint a figure, table, or other excerpt requires the consent of one of the original authors and notification to AIP. Republication or systematic or multiple reproduction of any material in this volume is permitted only under license from AIP. Address inquiries to Series Editor, AIP Conference Proceedings, AIP, 335 East 45th Street, New York, NY 10017-3483.

L.C. Catalog Card No. 91-58537
ISBN 0-88318-936-4
DOE CONF-9105255

Printed in the United States of America.

Joint Soviet-American Workshop on the Physics of Semiconductor Lasers May 20–June 3 1991

Contents

Foreword v
Introduction remarks vii
    Zh.I. Alferov

Session A: High power and low threshold lasers

Use of native oxides in $Al_xGa_{1-x}As$ QWH lasers 1
    R.D. Burnham, J.M. Dallesasse, S.C. Smith, N. Holonyak, Jr., F.A. Kish,
    A.R. Sugg, N. El-Zein, T.A. Richard, and S.J. Caracci

An experimental and theoretical study of the local temperature rise of mirror facets in InGaAsP/GaAs and AlGaAs/GaAs SCH SQW laser diodes 6
    D.Z. Garbuzov, N.I. Katsavets, A.V. Kochergin, and V.B. Khalfin

High-power phase-locked arrays of antiguides 14
    D. Botez

Low threshold quantum well AlGaAs-heterolasers fabricated by low temperature liquid phase epitaxy 24
    V.M. Andreev, A.B. Kazantsev, V.R. Larionov. V.D. Rumyantsev, and
    V.P. Khvostikov

Estimation of output power from semiconductor laser limited by optical nonlinearity† ... 33
    P.G. Eliseev and R.F. Nabiev

High-power grating tuned semiconductor diode lasers and single-frequency diode-pumped Nd:YAG microcavity lasers 37
    P. Gavrilovič, S. Singh, V.B. Smirnitskii, J. Bisberg, and M. O'Neil

The influence of leakage on the characteristics of QW lasers 49
    V.B. Khalfin, A.B. Gulakov, I.V. Kochnev, E.U. Rafailov, Yu.M. Shernyakov,
    B.S. Yavich, and D.Z. Garbuzov

Session B: Fast switching

Nonlinear effects in picosecond high-power diode lasers 58
    E.L. Portnoi, E.A. Avrutin, and A.V. Chelnokov

High frequency modulation of quantum well heterostructure diode lasers by carrier heating in microwave electric field 67
    S.A. Gurevich, I.I. Filatov, B.M. Gorbovitsky, and V.B. Gorfinkel

Active Q-switching of GaAlAs/GaAs lasers using free carrier effects in a modulation doped QW 75
    Yu.G. Kozlov, T.V. Shubina, A.A. Toropov, and I.Yu. Shvechikov

Degenerate sixwave mixing in the active region of a diode laser and a problem of lateral distribution stability‡ 85
    A.P. Bogatov

Session C: Visible and long-wavelength lasers

Blue semiconductor laser research at the University of Florida‡ 86
    P.S. Zory

Tunable diode lasers for 3 to 40 μm infrared spectral region 87
    A.P. Shotov

InGaSbAs/GaAlSbAs heterostructures for mid-infrared injection lasers‡ 95
    B.N. Sverdlov

Session D: Materials

Extremely smooth AlGaAs-GaAs quantum wells grown by metalorganic chemical vapor deposition 96
    R.D. Dupuis, J.G. Neff, and C.J. Pinzone

***In situ* patterning of impurity induced layer disordering and other applications of laser patterned desorption** 104
    J.E. Epler, R.D. Bringans, T.L. Paoli, and D.W. Treat

Neutral impurity disordering of III–V quantum well structures for optoelectronics 111
    J.H. Marsh, S.G. Ayling, A.C. Bryce, S.I. Hansen, and S.A Bradshaw

Reactive ion etching for fabrication of intergrated optic and optoelectronic elements 130
    F.N. Timofeev

Session E: Surface emitting, tunable and frequency stabilized lasers

Monolithic multiple wavelength tunable vertical cavity surface emitting laser array 151
    C.J. Chang-Hasnain, M.W. Maeda, J.P. Harbison, and L.T. Florez

Semiconductor microcavity effect on spontaneous emission 172
    D.G. Deppe, C. Lei, T.J. Rogers, and B.G. Streetman

Frequency stabilized diode lasers‡ 182
    V.L. Velichansky

Authors' index 183

---

† The authors failed to participate in the Workshop Sessions but they have represented the paper
‡ Presented by abstract only

Foreword

The first Soviet-American Workshop on the Physics of Semiconductor Lasers was held in the USSR from May 19 to June 3, 1991. It was convened in accordance with the agreement between the Academy of Sciences of the USSR and the National Academy of Sciences signed in late 1988, and was co-sponsored by these two Academies. The idea behind the Workshop was to assemble leading specialists in semiconductor lasers from the countries so as to offer them a possibility to discuss the latest results obtained in this challenging area of physics. The co-chairmen of the Workshop were Academician Zh.I. Alferov (A.F. Ioffe Physico-Technical Institute, Leningrad) and Professor R.D. Dupuis (University of Texas at Austin). The Ioffe Institute took upon itself the job of organizing this meeting.

The Workshop sessions were held in Leningrad (May 19–26), after which (May 27—June 3) its participants visited a number of research institutes and laboratories in Minsk (the capital of Byelorussia) and Moscow. Among the participants of the Workshop there were 7 scientists from the USA, 11 from the USSR, and one each from Great Britain and Switzerland, who gave one-hour talks. The scientific program reflected the most important area of research in the physics and technology of semiconductor lasers:
— high-power and low-threshold lasers;
— fast switching;
— visible and long-wavelength lasers;
— materials;
— surface emitting, tunable, and frequency-stabilized lasers.

The papers that were read at the Workshop are presented in these Proceedings in the order they were arranged in the program. Regrettably, some of the authors did not submit the texts of their papers. These papers are presented by abstracts only.

Joint Soviet-American Workshop on the Physics of Semiconductor Lasers   May 20–June 3 1991

Introduction Remarks

Research on semiconductor lasers in the USA and USSR has been intensively carried out since the day of the birth of this wonderful solid state device. As a matter of fact, semiconductor lasers owe their origin to the scientists of these two countries. Even today, almost thirty years later they attract attention of physicists and engineers at hundreds of laboratories all over the world not only due to their wide applications—from compact discs to fiber optical communications—but also to unique, by diversity and scale, physical phenomena. Everything that is characterized by the words "sub" and especially "super" is widely used in these devices—submicron size, superfast operation, superthin layers, superinjection, etc. Development of the physics of low-dimensional electron gas structures, radiative recombination and fast phenomena in semiconductors is very closely connected with the semiconductor laser research. The idea to hold joint Soviet-American workshops on semiconductor lasers arose long ago. It should be noted that in spite of the waves of cooling in state relations and limitations of the COCOM, scientific collaboration, exchange of ideas, mutual visits, active competition were going on. This once again confirms an international nature of science. However, scientific cooperation in this very sensitive area cannot but suffer from the political aspects.

The idea of a workshop was proposed in 1986 when a new agreement on cooperation between the USSR Academy of Sciences and the USA National Academy of Sciences was signed. Nevertheless, it was only this year that the Workshop has been held in Leningrad. However, in autumn 1989 an international semiconductor laser workshop has been held in a beautiful secluded place, an old monastery near Plovdiv, the House of Architects now, on the initiative of the Bulgarian Academy of Sciences. Its main participants were American and Soviet scientists, who had feeling of deep gratitude to its organizer, Dr. Kakanakov (Institute of Applied Physics, Bulgarian Academy of Sciences). It became an example for holding the Workshop in Leningrad. The choice of the Workshop program was determined not only by the hot points of the physics of semiconductor lasers but by personal tastes of the organizers as well, which reflect the influence of two research schools, one of the Ioffe Physico-Technical Institute and the other of Professor N. Holonyak of the University of Illinois. However, as they say in this country, personal interests reflect the society ones.

No doubt, new drastic ideas in physics and technology, such as transition from homo- to heterostructures in ternary and then in quaternary systems, MBE and MOCVD technology, quantum well structures belong to history. Today these ideas are being developed to create devices with unique characteristics and new promising fields of application. The most impressive results of the recent years are the dramatic increase of power and efficiency and speed of response. Some of these problems were discussed by D. Botez in the paper devoted to AlGaAs high-power arrays and by D. Garbuzov on high power InGaAsP SCH SQW laser diodes. E. Portnoi considered nonlinear effects in high-power picosecond diode lasers, and S. Gurevich dwelled on new methods of effective high frequency modulation of the diode lasers by carrier heating. A tradition to elect the best speaker by ballot was born in Plovdiv. In Leningrad the prize—a bottle of champaign—was awarded to C. Chang-Hasnain. Studies of the surface emitting lasers have recently acquired practical importance due to the progress of the superlattice mirror technology, and C. Chang-Hasnain's paper was unanimously honored with the highest appraisal both for its content and the way of presentation. P. Gavrilovič described high-power tunable operation of AlGaAs QW lasers. This paper reflects another important feature of the Workshop—since it is *joint* work done by Microelectronics

Laboratory, Polaroid Corporation and the Ioffe Institute. D. Deppe suggested a very elegant analogy to the quantum size effects in quantum wells—the influence of a small optical cavity on the photon emission properties in the SE semiconductor lasers. In the recent years long wavelength lasers have become increasingly important for various applications. These problems were discussed in the papers by B. Sverdlov on injection lasers on the base of $A^{III}B^{V}$ heterostructure solid solutions (InGaAsSb/GaAlSbAs) for mid-infrared spectral range (1.8–2.4 μm), and by A. Shotov on QW lasers with wavelength of 5–10 μm on the base of $A^{IV}B^{VI}$ compounds.

One of the surprises of the Workshop was the talk by R. Burnham reporting the data on native oxides in AlGaAs structures. Joint studies of the University of Illinois and Amoco Research Center resulted in invention of native oxides for AlGaAs heterostructures. The authors have already applied these oxides to stripe geometry laser technology. These studies look quite promising for the development of microelectronics on the base of AlGaAs heterostructures. Very interesting communications were presented by A. Toropov who reported some new results on active Q-switching of AlGaAs lasers using free carrier effects in a modulation-doped GaAs QW, and by F. Timofeev on a reactive ion etching process in AlGaAs. As always excellent (remember Plovdiv) were the papers by P. Zory and V. Velichansky, the latter receiving the second prize of the Workshop.

As a whole, the last but not least Soviet–American Workshop on the Physics of Semiconductor Lasers became a festive occasion for its participants, and we hope that the publication of the Proceedings will be of use for all specialists working in this field.

Zh.I. Alferov

Use of native oxides in $Al_xGa_{1-x}As$ QWH lasers

R.D. Burnham, J.M. Dallesasse,[a] and S.C. Smith
Amoco Technology Company, Amoco Research Center,
Naperville, Illinois 60566, USA

N. Holonyak, Jr., F.A. Kish,[b] A.R. Sugg, N. El-Zein, T.A. Richard, and S.J. Caracci
Electrical Engineering Research Laboratory, Center for Compound Semiconductor Microelectronics, and Materials Research Laboratory,
University of Illinois at Urbana-Champaign, Urbana, Illinois 61801, USA

Abstract. Data will be presented on the long-term degradation of high Al composition $Al_xGa_{1-x}As$-GaAs quantum well heterostructures (QWH's). The data suggest that these heterostructures deteriorate via hydrolyzation during long term exposure to normal environmental conditions. The hydrolysis develops at cleaved edges, cracks, and fissures, and at pinholes in cap layers. However, at elevated temperatures ($T \geq 400\,°C$) high Al composition $Al_xGa_{1-x}As$-GaAs QWH's can be converted into a dense transparent native oxide by reaction with H_2O vapor in a N_2 carrier gas. This oxide has many of the properties necessary for device fabrication; it possesses a low index of refraction, it successfully masks impurity diffusion (Zn and Si), and thus can serve as a mask for impurity induced layer disordering (IILD), and it acts as an electrical insulator. Data on a single-stripe $Al_xGa_{1-x}As$-GaAs, multiple stripe $Al_xGa_{1-x}As$-GaAs, and $Al_yGa_{1-y}As$-GaAs-$In_xGa_{1-x}As$ QWH laser diodes fabricated with a native oxide current confinement layer will be presented. Finally, data on low-threshold buried-heterostructure $Al_xGa_{1-x}As$-GaAs QWH lasers fabricated by using both IILD and the native oxide process will be presented.

From almost the beginning of III–V work the binary Al-bearing III–V's have been listed as unstable.[1] In order to circumvent this problem particularly in the case of AlAs, the layers have either been passivated using anodization,[2] buried in QWH's,[3] or replaced with high-composition ($x \sim 0.8$) $Al_xGa_{1-x}As$ layers.[4] As is well known[1] the instability of AlAs is due to the extremely reactive nature of the Al, particularly in a moist environment. In the case of the use of AlAs layers in QWH's it was determined from long-term (≥ 8 years) studies that degradation via hydrolysis of the AlAs layers in QWH's occurred because of exposure at cleaved edges and pinholes in the cap layers. In addition QWH's containing thicker ($> 0.4\,\mu m$) AlAs layers were found to be much less stable than material containing thinner ($\simeq 200$ Å) AlAs layers. In particular the deterioration of a p-n QWH crystal grown in April, 1982 was examined at different intervals (June 1985, June 1986, and Dec. 1989) and an estimate of the rate at which the decomposition proceeded along the plane was found to be $\sim 0.37\,\mu m$/day (136 μm/yr).[5] Long-term (2–12 years) exposure at room temperature was also studied[6] in the case of high-composition ($x \geq 0.8$) $Al_xGa_{1-x}As$. The hydrolysis was found to be significant for thicker ($\geq 0.1\,\mu m$) $Al_xGa_{1-x}As$ layers of higher composition ($x \geq 0.85$). A comparison of electron dispersion x-ray spectroscopy (EDS) analysis of intact and hydrolyzed regions of an $Al_xGa_{1-x}As$-AlAs-GaAs QWH provides additional insight into the degradation mechanism.[6] The data indicates that As is lost from the crystal most likely in the form of volatile As compounds such as AsH_3 and the remaining As may be in the form of

© 1992 American Institute of Physics

an oxide such as As_2O_3. The presence of oxygen in the EDS data suggests that the Al reacts to form some combination of the following oxygen-bearing forms of Al:AlO(OH), Al(OH)$_3$, Al_2O_3, $Al_2O_3 \cdot H_2O$, and $Al_2O_3 \cdot 3H_2O$.[7,8]

In an attempt to accelerate the hydrolyzation oxidation process Dallesasse and co-workers[9] raised the QWH crystal temperature and passed water vapor over it, the classic method (after Frosch, 1955) to oxidize and mask Si.[10] This process had not been recognized as particularly useful for III–V semiconductors but the experiments indicated that instead of destructive hydrolyzation, smooth dense native oxides on $(AlAs)_x(GaAs)_{1-x}$ and $Al_xGa_{1-x}As$ can be formed by hydrolyzation at $T \geq 400\,°C$ in a furnace supplied with a N_2 carrier gas bubbled through H_2O at a temperature of 95 °C. Again it was found that fine-scale alloys $AlAs(L_B)$-$GaAs(L_Z)$ SL ($L_B+L_Z \leq 100$ Å) or random alloy $Al_xGa_{1-x}As$, $x \geq 0.7$ were more stable than coarse-scale alloys ($L_B+L_Z \geq 200$ Å) to the oxidation process. The degree of coarseness (SL period) is directly related to the degree ("speed") of oxidation. Oxidation normal to the crystal surface was found to proceed more slowly than from a cleaved edge.

An important question which arises is the particular form of this native oxide. Initially, the water-vapor oxidation of $Al_xGa_{1-x}As$ was believed to result in the formation of Al_2O_3 or one of its hydrated forms, $Al_2O_3 \cdot nH_2O$ ($0 \leq n \leq 3$). Another possibility is that AlO(OH) may be the product of oxidation at 400 °C or at higher temperatures, γ-Al_2O_3 may be formed. Annealing at elevated temperature improves the resistance of the grown oxide to acids, suggesting that the oxide changes form, perhaps to α-Al_2O_3. In contrast, the crystal deterioration at room temperature (and even mildly elevated temperatures) is believed to result in the formation of Al(OH)$_3$. In subsequent work, these speculations will be tested. An initial examination of native oxide stability indicates that a $\sim 0.1\,\mu$m-thick uniform dark blue native oxide layer formed from the AlAs layer was shown to be stable with aging (~ 100 days), while unoxidized samples degrade to a yellowish orange color after several days, with penetration through the AlAs ($0.1\,\mu$m) down into the GaAs as deep as $\sim 1\,\mu$m.[11] Therefore, relative to oxides formed ($\sim 25\,°C$) on AlAs (or $Al_xGa_{1-x}As$, $x \geq 0.7$) under atmospheric conditions (hydrolysis), oxides formed (via $N_2 + H_2O$) at higher temperatures (400 °C) are much more stable and seal the underlying crystal (e.g., GaAs).

In an effort to study the electrical and optical properties of devices which incorporate the native oxide process single-stripe $Al_xGa_{1-x}As$-GaAs QWH lasers were fabricated.[12] The epitaxial layers for these single stripe QWH lasers were grown on n-type (100) GaAs substrates by metalorganic chemical vapor deposition (MOCVD).[13] A GaAs buffer layer was grown first, followed by an n-type $Al_{0.8}Ga_{0.2}As$ lower confining layer. The active region of the QWH was grown next and consisted of a ~ 400 Å GaAs QW with ~ 1000 Å, $Al_{0.25}Ga_{0.75}As$ waveguide layers (undoped) on either side. Finally, a p-type $Al_{0.8}Ga_{0.2}As$ confining layer (~ 9000 Å) was grown on top of the active region and then capped with a heavily doped p-type GaAs contact layer (~ 800 Å thick). Standard photolithography, plasma etching, and wet chemical etching were used to form $10\,\mu$m-wide SiO_2-GaAs stripe regions. The remaining surface had $Al_{0.8}Ga_{0.2}As$ exposed. A native oxide region was then formed from ~ 1500 Å of the exposed $Al_{0.8}Ga_{0.2}As$ layer by heating the QWH crystal at $\sim 400\,°C$ for 3 h in a $N_2 + H_2O$ vapor atmosphere. After the SiO_2 removal the wafer was further processed into laser diodes. The native-oxide single-stripe AlGaAs-GaAs QWH lasers had excellent spectral quality and operated to powers in excess of 100 mW per facet.

Native-oxide-defined coupled-stripe $Al_xGa_{1-x}As$-GaAs QWH lasers were also fabricated in essentially the same way as the single-stripe devices except the GaAs contact layer was used as the oxide mask instead of a SiO_2 layer.[14] This was possible because the GaAs layer does not oxidize readily. Arrays of ten $5\,\mu$m-wide emitters on $7\,\mu$m centers were optically coupled and operated at powers as high as 300 mW per facet, and with wider stripe spacing ($5\,\mu$m emitters on $10\,\mu$m centers) as high as 400 mW per facet.

Also, high performance native-oxide coupled-stripe $Al_yGa_{1-y}As$-GaAs-$In_xGa_{1-x}As$ QWH lasers have been studied.[15] Again the epitaxial layers for these coupled-stripe QWH lasers were grown on n-type GaAs substrates by MOCVD. The growth temperature was 800 °C except for the $In_xGa_{1-x}As$ QW which was grown at 630 °C to prevent In desorption. A GaAs buffer layer was grown first, followed by an n-type (Se) $Al_{0.8}Ga_{0.2}As$ lower confining layer ($\sim 1.25\,\mu m$), and then the active region and waveguide layers (~ 3000 Å). The nominally undoped active region and waveguide (WG) consist of an $Al_{0.3}Ga_{0.7}As$ WG on either side of GaAs (~ 640 Å) with an $In_{0.1}Ga_{0.9}As$ QW (~ 100 Å) at its center. Then a p-type (Mg) $Al_{0.8}Ga_{0.2}As$ upper confining layer (~ 6000 Å) was grown with a p-type (Mg + Zn) GaAs contact layer (~ 1000 Å) on top.

The coupled-stripe lasers were fabricated by etching the GaAs cap except in ten $5\,\mu m$ wide stripes with a center-to center spacing of $10\,\mu m$. The remaining portion of the surface had the $Al_{0.8}Ga_{0.2}As$ upper confining layer exposed between and outside of the stripes. The crystal was then place in a $N_2 + H_2O$ vapor atmosphere at 400 °C for 3 h in order to grow the native oxide. The wafer was then further processed and packaged for laser characterization. The ten stripe arrays have a CW threshold of 95 mA at room temperature and an emission wavelength of ~ 913 nm. The couple-stripe lasers have a differential quantum efficiency of 42 % (uncoated facets) and were capable of power output of over 350 mW per facet. By changing the spacing (5300 Å to 4000 Å) of the waveguide and quantum well (WG + QW) region from the oxide between the stripes, it has been shown that the 10-stripe laser operates decoupled because of increased losses and the onset of index guiding. This is due to the fact that the electric field extends past the waveguide region into the confining layer and oxide. The loss can be attributed to absorption in the oxide, at the interface states, and below the oxide where the crystal (~ 1000 Å) suffers some oxygen and hydrogen "contamination."[11]

By further investigations, it was found that these native-oxides can also serve as an effective mask against impurity diffusion[16,17] of Zn and Si. Both SL and QWH crystals were grown by MOCVD for these experiments. In the case of impurity induced layer disorder (IILD) by Zn diffusion, the SL crystal consisted of a GaAs buffer layer grown first and next an undoped $Al_{0.8}Ga_{0.2}As$ lower confining layer ($\sim 0.1\,\mu m$). Then the SL consisting of 40 GaAs wells ($L_Z \sim 110$ Å) and 41 $Al_{0.4}Ga_{0.6}As$ barriers ($L_B \sim 150$ Å) is grown. The total SL thickness is $\sim 1.05\,\mu m$. Finally a 1000 Å $Al_{0.8}Ga_{0.2}As$ confining layer is grown on top of the SL which is capped with 3000 Å of GaAs.

As for the IILD by Si diffusion one SL consisted of 20 GaAs wells ($L_Z \sim 500$ Å) and 21 $Al_{0.5}Ga_{0.5}As$ barriers ($L_B \sim 500$ Å) that were confined on the upper and lower sides with 1000 Å of $Al_{0.8}Ga_{0.2}As$. Another SL consisted of 15 GaAs wells ($L_Z \sim 335$ Å) and 16 $Al_{0.5}Ga_{0.5}As$ barriers ($L_B \sim 335$ Å) confined with ~ 1000 Å of $Al_{0.6}Ga_{0.4}As$ on both sides. The QWH crystal employed here consisted of an n-type GaAs buffer layer, an undoped ~ 2000 Å $Al_{0.25}Ga_{0.75}As$ waveguide region, a $\sim 0.9\,\mu m$ p-type $Al_{0.8}Ga_{0.2}As$ upper confining layer, and heavily doped p-type GaAs cap (~ 800 Å) layer on top. In the center of the waveguide region was an undoped 200 Å $Al_{0.06}Ga_{0.94}As$ QW.

Most recently, IILD along with oxidation (native oxide) of high-gap $Al_xGa_{1-x}As$ confining layers were employed to fabricate low-threshold stripe-geometry buried-heterostructure $Al_xGa_{1-x}As$-GaAs QWH lasers.[18] Again, the QWH laser crystal was grown by MOCVD on an n-type substrate. The growth consisted of n-type buffer layers of GaAs ($\sim 0.5\,\mu m$) and $Al_{0.25}Ga_{0.75}As$ ($\sim 1\,\mu m$). This was followed by the growth of a $\sim 1.1\,\mu m$ $Al_{0.77}Ga_{0.23}As$ n-type lower confining layer, a ~ 2000 Å $Al_{0.25}Ga_{0.75}As$ undoped waveguide region, a $\sim 1.1\,\mu m$ $Al_{0.8}Ga_{0.2}As$ p-type upper confining layer, and a $\sim 0.1\,\mu m$ p-type GaAs cap. The center of the waveguide consisted of an ~ 200 Å $Al_{0.06}Ga_{0.94}As$ quantum well (undoped).

The laser diode fabrication process begins with a shallow Zn diffusion over the entire surface in an evacuated quartz ampoule (540 °C, 30 min). The shallow p^+ layer formed by the diffusion helps control lateral Si diffusion at the crystal surface under the masked regions in later processing steps.[18] After Zn diffusion Si_3N_4 stripes were formed by a sequence of chemical vapor deposition (CVD), photolithography and etching. Then CVD is used again to deposit a ~ 300 Å Si layer (550 °C) and a ~ 1700 Å SiO_2 cap layer (400 °C). Next the crystal is sealed in an evacuated quartz ampoule and annealed with excess As at 850 °C for 6.5 h. The high-temperature anneal results in Si diffusion and IILD outside of the GaAs contact stripes. All encapsulants were then removed by etching with a CF_4 plasma, and a ~ 2000 Å native oxide was grown by oxidation of the exposed high-gap upper confining layers resulting in self-aligned contact stripes. Another shallow Zn diffusion had to be given on the contact stripes prior to metallization and packaging. QWH lasers fabricated by this method have continuous 300 K threshold currents as low as 5 mA and powers 31 mW per facet for ~ 3 μm wide active regions.

In conclusion, data has been presented on the properties of atmospherically hydrolyzed and water-vapor oxidized (native oxide: $N_2 + H_2O$, 400 °C) $Al_xGa_{1-x}As$-GaAs ($x \geq 0.75$) heterostructures and AlAs-GaAs superlattices. It has been demonstrated that a variety of single and multiple stripe $Al_xGa_{1-x}As$ based QWH laser devices can be fabricated using the native oxide as part of the processing procedure. In addition, the native oxide is an effective means of reducing leakage currents in the diffused IILD shunt junctions. Additionally, the oxidation process results in a self-aligned structure that is independent of the disordering techniques employed, relying only on the presence of exposed high-gap $Al_xGa_{1-x}As$ outside the contact stripe. As these oxidation processes become more sophisticated, lower composition $Al_xGa_{1-x}As$ and other Al-bearing III–V compounds may also be transformed into high-quality oxide, resulting in the increased generality of these techniques to other types of laser crystals.

Acknowledgments

The authors are grateful to R.W. Kaliski, C.D. Dickinson, J.T. Niccum, M.L. Savel and R.Y. De-Jule (Amoco Technology Company); K.C. Hsieh, D.C. Hall, G.E. Hofler, J.E. Baker, E.J. Vesely, M.J. Ries, R.T. Gladdin and B.L. Payne (University of Illinois); P. Gavrilovic, K. Meehan and J.E. Williams (Polaroid Corporation); R.D. Dupuis (University of Texas at Austin). Work at the University of Illinois was supported by the Army Research Office Contract No. DAAL-03-89-K-0008, National Science Foundation Grants No. DMR 89-20538 and No. ECD89-43166.

References

[a] Work performed under N. Holonyak, Jr., Electrical Engineering Research Laboratory, Center for Compound Semiconductor Microelectronics, and Materials Research Laboratory, University of Illinois, Urbana, Illinois, 61801.
[b] AT and T Doctoral Fellow.
[1] C. Hilsum and A.C. Rose-Innes 1961 *Semiconducting III–V Compounds* (Pergamon, Oxford), p. 3
[2] W.D. Johnson Jr. and W.M. Callahan 1976 *Appl. Phys. Lett.* **28** 150
[3] W.D. Laidig, N. Holonyak Jr., M.D. Camras, K. Hess, J.J. Coleman, P.D. Dapkus, and J. Bardeen 1981 *Appl. Phys. Lett.* **38** 776
[4] R.D. Dupuis (private communication). Dupuis attempted AlAs windows on GaAs solar cells grown by metalorganic chemical vapor deposition but abandoned their use immediately upon noticing their instability (changing color in room air). See R.D. Dupuis, P.D. Dapkus, R.D. Yingling, and L.A. Moudy 1977 *Appl. Phys. Lett.* **31** 201

5. J.M. Dallesasse, P. Gavrilovic, N. Holonyak, Jr., R.W. Kaliski, D.W. Nam, E.J. Vesely, and R.D. Burnham 1990 *Appl. Phys. Lett.* **56** 2436
6. J.M. Dallesasse, N. El-Zein, N. Holonyak, Jr., K.C. Hsieh, R.D. Burnham and R.D. DuPuis 1990 *J. Appl. Phys.* **68** 2235
7. M. Pourbaix 1973 *Lectures on Electrochemical Corrosion* (Plenum, New York), pp. 145
8. M. Pourbaix 1966 *Atlas of Electrochemical Equilibria in Aqueous Solutions* (Pergamon, London), pp. 168–176 and 516
9. J.M. Dallesasse, N. Holonyak, Jr., A.R. Sugg, T.A. Richard, and N. El-Zein 1990 *Appl. Phys. Lett.* **57** 2844
10. C.J. Frosch and L. Derick 1957 *J. Electrochem. Soc.* **104** 547
    See also C.J. Frosch 1958 *Transistor Technology* **3** 90
11. A.R. Sugg, N. Holonyak, Jr., J.E. Baker, F.A. Kish, and J.M. Dallesasse 1991 *Appl. Phys. Lett.* **58** 1199
12. J.M. Dallesasse and N. Holonyak, Jr. 1991 *Appl. Phys. Lett.* **58** 394
13. R.D. Dupuis, L.A. Moudy and P.D. Dapkus 1979 *Proceedings of 7th International Symposium on GaAs and Related Compounds* Institute of Physics, London, pp. 1
14. J.M. Dallesasse, N. Holonyak, Jr., D.C. Hall, N. El-Zein, A.R. Sugg, S.C. Smith, and R.D. Burnham 1991 *Appl. Phys. Lett.* **58** 834
15. T.A. Richard, F.A. Kish, N. Holonyak, Jr., J.M. Dallesasse, K.C. Hsieh, M.J. Ries, P. Gavrilovic, K. Meehan, and J.E. Williams 1991 *Appl. Phys. Lett.* **58** (to be published)
16. J.M. Dallesasse, N. Holonyak, Jr., N. El-Zein, T.A. Richard, F.A. Kish, A.R. Sugg, R.D. Burnham, and S.C. Smith 1991 *Appl. Phys. Lett.* **58** 974
17. N. El-Zein, N. Holonyak, Jr., F.A. Kish, A.R. Sugg, T.A. Richard, J.M. Dallesasse, S.C. Smith, and R.D. Burnham *J. Appl. Phys.* (to be published)
18. F.A. Kish, S.J. Caracci, N. Holonyak, Jr., J.M. Dallesasse, G.E. Hofler, R.D. Burnham, and S.C. Smith 1991 *Appl. Phys. Lett.* **58** 1765

Joint Soviet-American Workshop on the Physics of Semiconductor Lasers May 20–June 3 1991

An experimental and theoretical study of the local temperature rise of mirror facets in InGaAsP/GaAs and AlGaAs/GaAs SCH SQW laser diodes

D.Z. Garbuzov, N.I. Katsavets, A.V. Kochergin, and V.B. Khalfin

A.F. Ioffe Physico-Technical Institute, Academy of Sciences of the USSR,
26 Polytekhnicheskaya st. 194021 Leningrad, USSR

Abstract. A photoluminescence study has been carried out of the local mirror facet temperature rise for InGaAsP/GaAs and AlGaAs/GaAs SCH SQW high power broad contact laser diodes. It is shown that the local optical temperature rise of mirror facets (ΔT_{opt}^m) in conventional AlGaAs/GaAs SCH SQW laser diodes is nearly 5–10 times higher than that in InGaAsP/GaAs lasers operating under the same conditions. It has been established that ΔT_{opt}^m decreases with increasing waveguide width (d_w), and that for InGaAsP/GaAs diodes with $d_w = 0.8$ μm, ΔT_{opt}^m may be not in excess of 5° for uncoated facet diodes with a 100 μm aperture for output power of 1 W. Calculations have shown that the surface recombination model cannot account for the experimental data on mirror facet overheating in lasers with a single quantum well active region. To explain these results, one has to assume the existence in the active region of a near-facet dead layer of finite thickness with an enhanced nonradiative recombination rate.

Broad waveguide lasers characterized by a low local mirror facet temperature rise represent the most promising modification of laser diodes permitting one to increase their maximum output power density.

The present communication reports on a study of the local temperature rise of mirror facets in InGaAsP/GaAs and AlGaAs/GaAs lasers and a comparison of the results obtained with calculations.

The temperature of the AlGaAs/GaAs laser diode mirror facets has been a subject of investigation in a number of preceding studies.[1-4] These studies have led to a common conclusion that the mirror facet local temperature rise grows superlinearly with output power and revealed that the temperature rise values for uncoated facets in AlGaAs/GaAs lasers exceed 100 °C at output power densities of about 1 MW/cm^2.

In most of the studies, the local temperature rise of the facets was interpreted in terms of the surface recombination model[1,2] by which nonradiative recombination of nonequilibrium carriers diffusing along the active layer occurs only at the facet surface while the lasing mode absorption and nonequilibrium carrier generation takes place in the region of reduced concentration of these carriers adjoining the facet surface, whose dimension $\mathcal{L}$ does not exceed the ambipolar diffusion length ($\mathcal{L} < \mathcal{L}_d$). Within this model, the superlinear behaviour of the $\Delta T_{opt}^m = f(P)$ dependence is explained as due to an enhancement of absorption caused by a narrowing of the energy gap in the overheated near facet region of the active layer. In our report submitted to CLEO-91[5] we stressed that studies of the local temperature rise of facets in AlGaAs/GaAs SCH SQW lasers cast doubt on the validity of this model and show that at least for thin active layer structures ($L_z \simeq 100$ Å), calculations made in the frame of this model are not capable of providing even a qualitative explanation for the available experimental data. The same report[5] presented the first measurements of the local facet temperature rise in InGaAsP/GaAs SCH SQW laser diodes emitting in the same wavelength range ($\lambda = 0.8$ μm) as the AlGaAs/GaAs lasers and having close light output-current characteristics.

The present paper provides for the first time a detailed description of the experimental techniques used, gives new results, and compares them with theoretical calculations.

The subject of the study was broad contact area AlGaAs/GaAs and InGaAsP/GaAs SCH SQW 100 μm-stripe laser diodes. The AlGaAs/GaAs and InGaAsP/GaAs structures used to fabricate these diodes were grown by MOCVD and modified LPE.[6,7] Fig. 1 presents the band diagrams of the structures specifying the layer thickness. In the case of InGaAsP/GaAs structures, besides samples with usual waveguide layer thickness ($d_w = 0.4$ μm), broader (0.8–1 μm) waveguide (SCBW SQW) structures were prepared and studied.

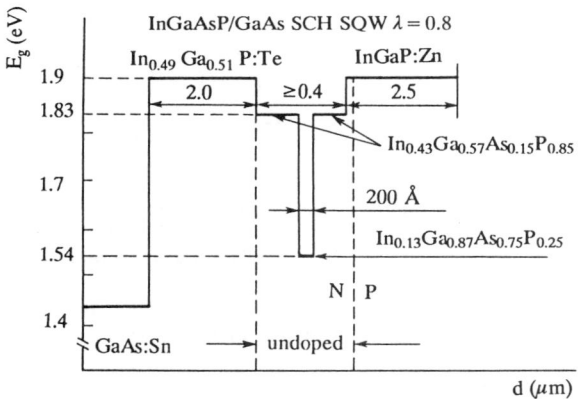

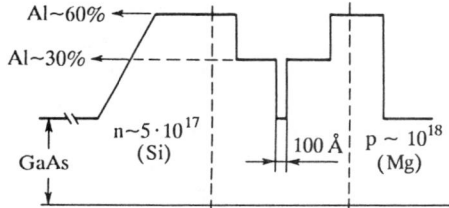

Fig. 1. Band diagrams of InGaAsP/GaAs and AlGaAs/GaAs laser structures.

Some of the studied diodes were prepared using a photolithographic mask permitting one to fabricate lasers whose stripe contact to the p-cladding had a 10×50 μm window. After the fabrication, the AlGaAs/GaAs structures with a stripe contact window were etched selectively to remove the GaAs contact layer from the window area.

The data presented below relate to diodes bonded p-cladding up. The diode output facets were uncoated.

The local temperature rise of the mirror facets was determined by the photoluminescence technique. The pump light was focused either onto the stripe window or onto the output mirror facet of the laser diode bonded to the copper heat sink. The radiation of the pump Ar$^+$ laser (488 nm) could be focussed by a microscope optical system into a spot of minimum diameter 3 μm. The luminescence radiation was collected by the same optical system and directed into a double MDR-23 spectrometer arrangement (Fig. 2).

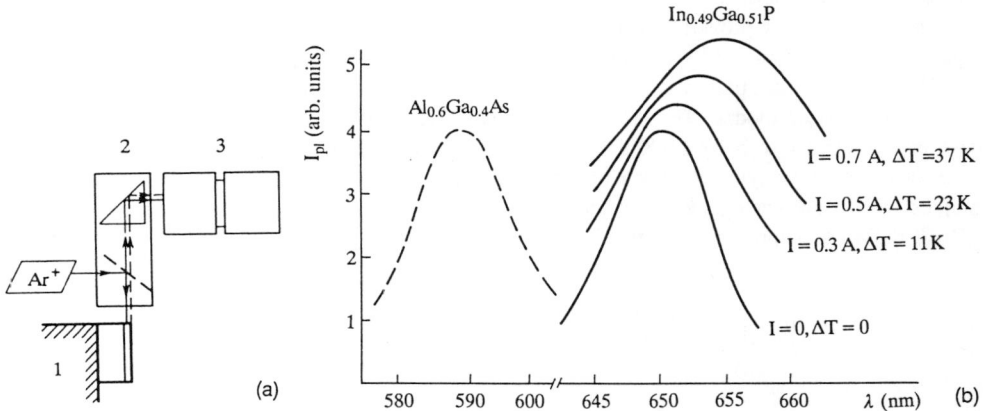

Fig. 2. (a) Schematic of the laser diode geometry in a photoluminescence experiment to determine the local temperature rise of the active region: (1) heat sink with laser diode; (2) microscope; (3) double spectrometer. (b) Photoluminescence spectra of $Al_{0.6}Ga_{0.4}As$ and $In_{0.49}Ga_{0.51}P$ cladding. Adjoining each spectrum is the value of the corresponding driving current through the diode.

The temperature was determined using the cladding luminescence peak lying far enough from the diode lasing line to ensure peak position reliable measurement even at laser power outputs above 1 W.

When pumping through the stripe contact window the position of the cladding peak characterized the temperature in the bulk of the active region, and when focusing the Ar$^+$ laser beam on the facet mirror, the temperature measured corresponds to that of the active region layer close to the mirror facet.

Calculations show that when considering the local temperature rise in the active region associated with heat liberation throughout the stripe area, the temperature difference between the p-cladding and the active region should not be in excess of 10%. With heat released in the active region only near facet the temperature gradients over the facet surface depend on the thickness of the near-facet region where the heat is released. In the limiting case of an infinitely thin heating region ($W \times L_z$ plane area), the difference between the temperature of this region and the mean cladding temperature (ΔT) may be as high as 50 %. If the source of heat is a layer of finite thickness $\mathcal{L}$ of size $W \times L_z \times \mathcal{L}$, this temperature difference decreases with increasing $\mathcal{L}$, so that for $\mathcal{L} = 1$ μm, ΔT already does not exceed 10–20%.

Fig. 2 presents luminescence spectra of an $In_{0.49}Ga_{0.51}P$ cladding layer obtained at the photoexcitation of the facet of an InGaAsP/GaAs laser diode operating in the CW mode at different currents as well as a photoluminescence spectrum of an $Al_{0.6}Ga_{0.4}As$ p-cladding of an AlGaAs/GaAs laser diode. As the driving current is increased, the photoluminescence intensity of the edge emission peak for the direct band-gap $In_{0.49}Ga_{0.51}P$ cladding layer decreases and its position shifts longward due to the temperature rise.

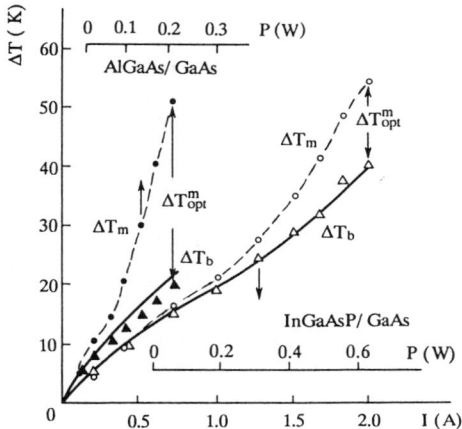

Fig. 3. The current dependence of the local temperature rise for the bulk of the active region and the output mirror facet typical of AlGaAs/GaAs SCH SQW lasers (filled data points) and of InGaAsP/GaAs lasers with a waveguide $\simeq 0.4\,\mu$m wide (unfilled data points). The triangles identify the local temperature rise of the bulk of the active region, and circles—the measured values for the mirror facets. Solid curves were obtained by fitting the experimental data to $\Delta T_b = R_T(IU - P)$ relations, with $R_T = 15\ °$/W for AlGaAs/GaAs diode and $R_T = 9\ °$/W for InGaAsP/GaAs diode. Plotted on the additional horizontal axes is CW power at the output mirror facet for AlGaAs/GaAs diode (upper axis) and for InGaAsP/GaAs diode (lower axis)

The edge emission luminescence intensity of the indirect-band gap $Al_{0.6}Ga_{0.4}As$ p-cladding was three orders of magnitude lower and increased with temperature, which suggests that the states involved in the corresponding transitions belong to the Γ-valley of the conduction band and that these transitions are of the direct nature. To reduce the local heating of the region under study by the pump light, the spot used was of a comparatively large diameter ($\approx 20\ \mu$m), so that the associated overheating did not exceed 2 °C for an absorbed light power of ~ 5 mW. Note that the dimensions of the region in question along the temperature gradient (perpendicular to the plane of the structure) were determined by the thickness of the cladding layers.

The calibration graphs, $h\nu_{max} = f(T)$, were constructed basing on special experiments in which the diode temperature was varied by heating properly the heat sinks. In these experiments calibration curves for the temperature-induced shift of the cladding peak were determined by comparising with that of edge emission peak for the n-GaAs substrate, for which the $h\nu_{max} = f(T)$ dependence is well known. (The luminescence of n-GaAs was excited in the substrate region adjoining the cladding layer through absorption of photons from the corresponding part of the pump spot.)

Fig. 3 presents an experimental dependence of temperature rise on diode current typical of the AlGaAs/GaAs SCH SQW lasers studied with standard layer thicknesses: $L_z = 100$ Å, $d_w = 0.2\ \mu$m, and for InGaAsP/GaAs diodes with $L_z = 200$ Å and $d_w = 0.4\ \mu$m. The unfilled and filled triangles in this figure specify the measured local temperature rise values for the bulk of the active region obtained when focussing the laser beam onto the stripe window, and the unfilled and filled circles, those for the laser diode mirror facets. The experimental data on the local temperature rise in the bulk of the active region can be fitted by the relation $\Delta T_b = R_T(IU - P)$, where P is the laser diode output power for the driving current I and applied bias U, and R_T is a constant whose value for most diodes bonded substrate down exceeds the calculated values of the thermal resistances by not more than 25%. The solid curves in Fig. 3 illustrate the calculations of ΔT_b obtained using this relation.

We are going to discuss now the difference between the dashed and solid curves in Fig. 3 corresponding to the local optical temperature rise of the laser diode mirror facet (ΔT_{opt}^m).

The values of ΔT_{opt}^m for the AlGaAs/GaAs laser diodes determined by the above mentioned method ($\Delta T_{opt}^m = 30\ °C$ for 0.25 MW/cm^2) are in a good agreement with available data obtained for similar devices by other techniques.[4] The values of ΔT_{opt}^m for the InGaAsP/GaAs laser diodes are nearly an order of magnitude smaller although the optical confinement factor (and, hence, the fraction of the lasing mode propagating along the active region) in these lasers is approximately equal to that for the AlGaAs/GaAs devices.

In an attempt to reduce ΔT_{opt}^m still more, a series of InGaAsP/GaAs structures with thicker waveguide layers have been recently prepared (SCBW SQW structures). Fig. 4(a) shows the dependence of ΔT_{opt}^m on the driving current and on optical output power typical of diodes with a waveguide thickness of 0.8–1 μm. As seen from a comparison of Figs 3 and 4(a), increasing the waveguide thickness does indeed reduce the local temperature rise of the facets which indicates that the corresponding mechanisms of heat release are connected with the lasing mode absorption directly in the active region.

Fig. 4(b) presents also the values of ΔT_{opt}^m characteristic of standard AlGaAs/GaAs SCH SQW diodes and compares them with the theoretical relations $\Delta T_{opt}^m = f(P)$ obtained in calculations for two different models of heat liberation near the mirror facet.

In the calculation, where the mirror facet temperature is related to heat source rate, we employed, irrespective of the model adopted, the same expressions derived under the following simplifying assumptions:
(i) the effects associated with the finite stripe width (W) were neglected, the analysis being reduced to the problem of an infinitely broad laser diode with linear radiation density $\simeq P/W$;
(ii) close to the heat source, the real geometry was replaced by the conditions corresponding to cylindrical symmetry, namely, the heat source was considered to be a halfcylinder of radius r; the solid solution layers of low thermal conductivity (λ_c) were substituted by a cylindrical layer of radius D equal to the total thickness of the layers of solid solutions separating the active region from the GaAs substrate, a material of high thermal conductivity (Fig. 4(c)).
(iii) heat was assumed to be removed only through the heat sink providing a constant temperature in the substrate plane at a distance b from the heat source (Fig. 4(c)).

Within these assumptions, Laplace's equation for the temperature distribution was solved numerically, and ΔT_{opt}^m was determined as a function of the heat source rate. An analysis shows that the numerical calculations of this relation can be fitted within a 50% accuracy by the following expression

$$\Delta T_{opt}^m = \frac{\Theta}{\pi}\left(\frac{1}{\lambda_c}\ln\frac{D}{\mathcal{L}} + \frac{1}{\lambda_s}\ln\frac{b}{D}\right) \qquad (1)$$

where Θ is the linear heat source power density (W/cm), λ_c is the thermal conductivity of the solid solution layers which, irrespective of their composition, was assumed to be one fourth that of the substrate (λ_s). The determination of the quantity r depends on the actual nonradiative recombination model chosen and will be considered later. For the case of interest to us here of heating

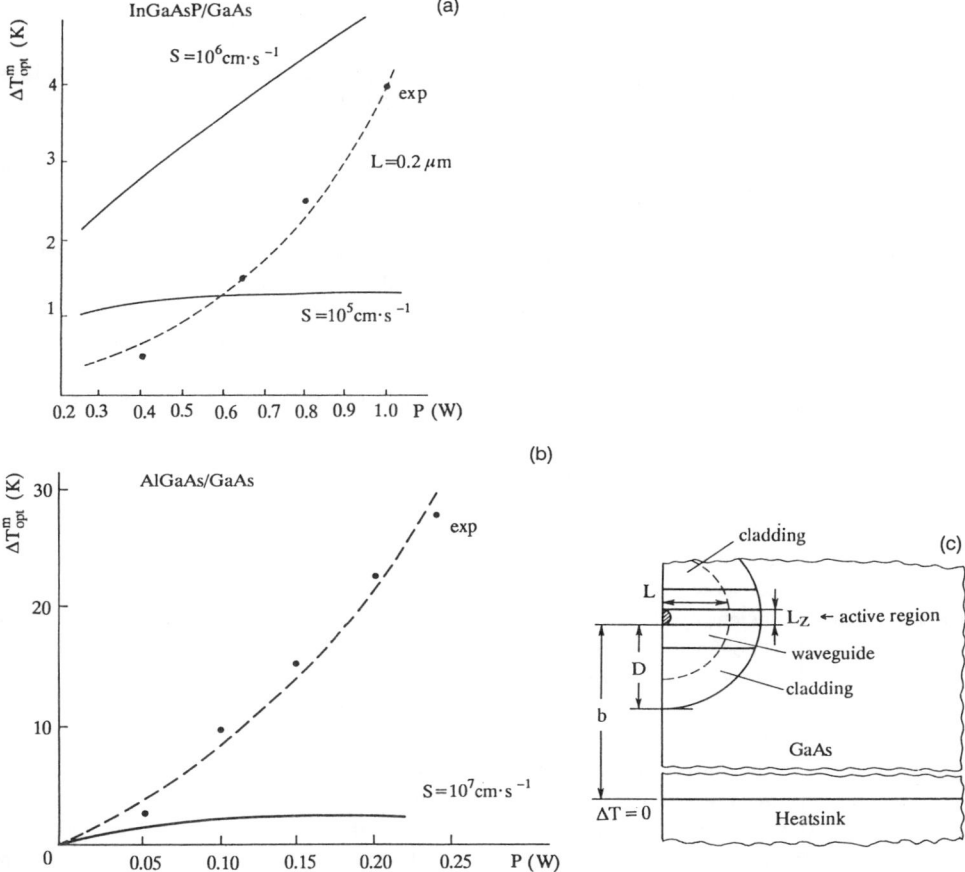

Fig. 4. (*a,b*) Dots—experimental values of ΔT_{opt}^m as a function of driving current for the AlGaAs/GaAs diode relating to Fig. 3, and for a InGaAsP/GaAs diode with a waveguide about 0.8 μm wide. Solid curves—calculation of ΔT_{opt}^m by the surface recombination model. The values of the surface recombination velocity are specified for each curve. The values of the other parameters are as follows: threshold nonequilibrium carrier concentration $n_{th} = p_{th} = 2 \times 10^{18}$ cm^{-3}. Ambipolar diffusion coefficient $D_{e-h} = 10$ cm^2/s. Dashed curves—calculation for the model of a dead layer of constant thickness $\mathcal{L}$. The values of the optical parameters used in the two models: InGaAsP/GaAs diode: $k_0 = 10^3$ cm^{-1}; $\partial k_0/\partial h\nu = 10^3$ cm^{-1}meV^{-1}; AlGaAs/GaAs diode: $k_0 = 7 \times 10^3$ cm^{-1}; $\partial k_0/\partial h\nu = 4.4 \times 10^2$ cm^{-1}meV^{-1}; (c) Schematic of the diode model used in calculations of the local mirror facet temperature rise.

associated with the lasing mode absorption in the active region, one can use the following relation to calculate Θ:

$$\Theta = \frac{AP}{W} \times \Gamma \times k_T^0 \times \mathcal{L} \quad (2)$$

where A is a coefficient determining the relation between the output power P and the lasing mode power inside the diode close to the output mirror facet ($A \simeq 2$ for an uncoated facet with $R = 0.3$),

$\mathcal{L}$ is the thickness of the absorbing layer in the active region adjoining the facet, k_T^o is the coefficient of the lasing mode absorption in the absorbing material which depends on its temperature rise and Γ–optical confinement factor for lasing mode.

The difference between the two theoretical models which are used in the calculations illustrated by Fig. 4(a,b) reduces to the difference in determining the quantity Θ. In the framework of the surface recombination model the dimensions of the region of heat liberation are much smaller than $\mathcal{L}$, and one can reasonably assume the value of r in relation (1) to be an the order of the active region thickness, $r \simeq L_z$.

As for the quantity $\mathcal{L}$, in the surface recombination model it is not constant but depends on laser output power. Indeed, it is obvious that the thickness of the near-surface layer (where the nonequilibrium carrier density is less than the threshold density) should decrease with increasing rate of nonequilibrium electron and hole generation in it.[1] As a result, the mirror facet heating should slow down with increasing power output approaching a constant level. An opposite, accelerating effect on the dependence $\Delta T_{opt}^m = f(P)$ is exerted by the local temperature rise in the near-surface region which increases the effective coefficient of absorption of the lasing mode in it:

$$k_T^0 = k_0 + \frac{\partial k_0}{\partial h\nu} \times \frac{\partial E_g}{\partial T} \times \frac{\partial T}{\partial \Theta} \times \Theta \tag{3}$$

The quantity $\partial T/\partial \Theta$ in (3) is found from (1), and the factor $\partial k/\partial h\nu$ can be evaluated, for instance, using the data from.[8]

The calculations show that the increase of ΔT_{opt}^m caused by the temperature-induced rise of the absorption coefficient can become significant only if ΔT_{opt}^m is greater than 30–40 °C. Since, however, the optical confinement factor Γ in SQW lasers is small, the growth of ΔT_{opt}^m should saturate due to the decrease of $\mathcal{L}$ even in the case of the maximum surface recombination velocities ($S = 10^7$ cm/s) already for $\Delta T_{opt}^m = 2-3°C$ (solid curve in Fig. 4(b)). Similar disagreement occurs when attempting to interpret the experimental relationships $\Delta T_{opt}^m = f(P)$ for InGaAsP/GaAs diodes in terms of the surface recombination model (solid curves in Fig. 4(a)).

In connection with this, in the second version of our calculations we assumed the nonradiative recombination to occur not on the mirror surface but rather throughout all the near-surface region, its thickness being determined by the distribution of additional near-surface nonradiative recombination centers and not depending on the output power. Obviously, under such an assumption the radius of the heat source r in (1) should be adopted equal to $\mathcal{L}$ (Fig. 4(c)).

One can now readily choose the parameters $\mathcal{L}$, k_0 and $\partial k_0/\partial h\nu$ providing the best fit between the experimental and calculated relations $\Delta T_{opt}^m = f(P)$, as this is shown in Fig. 4(a,b). The values of the parameters k_0 and $\partial k_0/\partial h\nu$ specified in the captions of figures were taken the same in the calculations by the two models and do not contradict the data of.[8] According to these calculations, the thickness of the dead layer in AlGaAs/GaAs lasers should be about 1 μm, and in InGaAsP/GaAs diodes, the value of $\mathcal{L}$ can be five times smaller.

The smaller effect of surface recombination in InGaAsP compared to GaAs and AlGaAs is confirmed by a direct comparison of the efficiency of luminescence under photoexcitation of the surface of the corresponding layers.

Thus, the main results of the present work can be summarized as follows:
(i) Experiments with InGaAsP/GaAs lasers show that the local temperature rise in laser diode mirror facets, in accordance with theoretical estimates, should decrease with decreasing optical confinement factor.
(ii) For SQW lasers characterized by small values of Γ the widely accepted model of interface recombination is not capable of accounting for the experimentally observed local temperature rises.

(iii) For a given value of the Γ factor the local temperature rise of mirror facets in InGaAsP/GaAs laser diodes is nearly an order of magnitude lower than that in AlGaAs/GaAs lasers due to the weaker influence of non-radiative recombination in the near-surface dead layer.

(iv) In broad waveguide InGaAsP/GaAs lasers local temperature rises of uncoated mirror facets of less than 5° have been demonstrated for a diode radiation density $\simeq 1$ MW/cm^2. Reduction of local facet temperature rise by increasing the wavequide width appears to be a promising way of increasing the limiting output power densities in laser diodes.

References

[1] C.H. Henry, P.M. Petroff, R.A. Logan and F.M. Merrit 1979 *J. Appl. Phys.* **50** 3721
[2] F. Kappeler, K. Mettler and K.H. Zschauer 1982 *IEE Proc.-1* **129** 256
[3] S. Todorski, M. Sawai and K. Aik 1985 *J. Appl. Phys.* **58** (3) 1124
[4] H. Brugger and P.W. Eppeillich 1990 *Appl. Phys. Lett* **56** 1049
[5] D.Z. Garbuzov, N.I. Katsavets, A.V. Kochergin, A.V. Michailov, E.U. Rafailov and V.B. Khalfin 1991 *Conf. on Lasers and Electro-Opt.* Baltimore, USA p. 142
[6] Zh.I. Alferov, D.Z. Garbuzov, S.N. Zhigulin, I.A. Kuz'min, B.B. Orlov, M.A. Sinitsyn, N.A. Strugov, V.E. Tokranov and B.S. Yavich 1988 *Sov. Phys.-Semicond.* **22** (12) 1334
[7] D.Z. Garbuzov, N.Yu. Antoniskis, A.D. Bondarev, A.B. Gulakov, S.W. Zhigulin, N.I. Katsavets, A.V. Kochergin and E.U. Rafailov 1991 *IEEE J. Quantum Electron.* **QE-27** (6)
[8] G. Lengyel, Kevin W. Jelly, Reinhurt W.H.Engelman 1990 *IEEE J. Quantum Electron.* **QE-26** (2) 286

Joint Soviet-American Workshop on the Physics of Semiconductor Lasers   May 20–June 3 1991

High-power phase-locked arrays of antiguides

D. Botez
TRW Research Center, Redondo, CA 90278, USA

Abstract. Phase-locked arrays of antiguides are a unique class of (monolithic) coherent diode lasers in that they provide both strong overall interelement coupling as well as strong optical-mode confinement. The leaky-wave characteristics of antiguides allow for a resonant condition; that is, when the interelement regions are odd integer numbers of the leaky-wave (lateral) half wavelength all elements equally couple to each other creating so called parallel coupling. By contrast, for the vast majority of coherent arrays published to date interelement coupling is of the nearest-neighbor type, so called series coupling, which gives weak coherence and poor intermodal discrimination. Parallel-coupled arrays of antiguides are called resonant-optical-waveguide (ROW) arrays. ROW devices posses such desirable properties as: full coherence, uniform intensity profile, and large intermodal discrimination. Thus, high coherent powers can be achieved without active phase control.
The theory of operation for ROW arrays will be outlined.
Experimental results include diffraction-limited-operation from both 20- and 40-element devices to high drive levels (10 × threshold) and powers (0.5–1.5 W). CW diffraction-limited operation has been achieved to 0.5 W, while in pulsed operation up to 2 W is obtained in a nearly diffraction-limited beam. Devices can be driven to 5 W with beamwidths 3 × diffraction limit.

Monolithic phase-locked arrays of GaAs/AlGaAs diode lasers, with no need for active phase control, are sought as reliable sources of high coherent powers (> 100 mW diffraction limited) for applications such as space communications, blue-light generation via frequency doubling, optical interconnects and parallel optical-signal processing. Conventional narrow-stripe ($3 - 4\,\mu$m wide) single-mode lasers mode provide at most 100 mW reliably, as limited by the optical power density at the laser facet. For reliable operation at watt-range power levels large-aperture ($\geq 100\,\mu$m) sources are necessary. Thus, the challenge has been to maintain a single spatial mode, from large-aperture devices, to high power levels. For this reason, phase-locked arrays have been under development for over a decade as a means of obtaining mode-stabilized devices operating reliably at powers in the 0.5–1.0 W range.

Monolithic (linear) arrays consisting of phase-locked index-guided lasers have proved useful tools in understanding the complex modal behavior of array devices. Positive-index-guided coupled oscillators, designed to operate in-phase, generally operate in several array modes above 1.5 × laser threshold or 50 mW.[1-3] None of the positive-index-guided monolithic laser arrays reported to date have demonstrated diffraction-limited-beam operation beyond $\cong$ 50 mW because of either weak overall coupling or weak optical-mode confinement. The real problem is that researchers have taken for granted that strong nearest neighbor coupling implies strong overall coupling. In reality, as shown in Fig. 1, nearest-neighbor coupling is "series coupling", a scheme plagued by weak overall coherence and poor intermodal discrimination. Strong overall coupling happens only when each element equally couples to all others, so called "parallel coupling". In turn intermodal discrimination is maximized and full coherence becomes a system characteristic. That is, there is virtually no reason to couple resonators unless parallel coupling can be achieved.

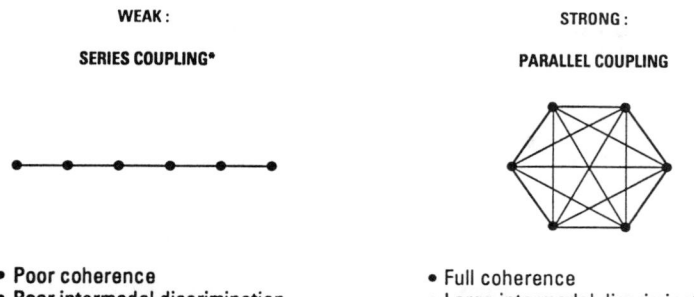

Fig. 1. Types of overall interelement coupling in phase-locked arrays

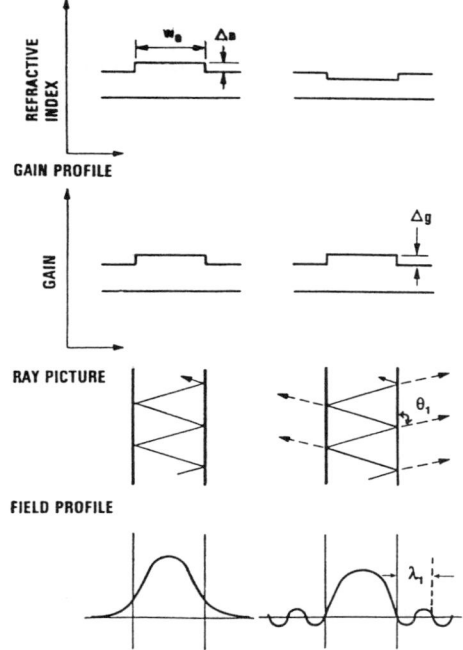

Fig. 2. Comparison between positive-index guides (left column) and negative-index guides (right column)

Phase-locked arrays of negative-index guides or antiguides is a new class of arrays, that possesses the unique feature of having both strong optical-mode confinement (within each element) and strong overall interelement coupling (i.e. parallel coupling via leaky waves). A single antiguide can be understood by comparison to a single positive-index guide (Fig. 2). While in a positive-index guide radiation is trapped via total internal reflection, in an antiguide radiation is only partially reflected at the antiguide-core boundaries. Light refracted into the cladding layers is radiation leaking outwardly with a lateral wavelength λ_1. It can be thought of as radiation loss, α_R.

$$\lambda_1 \cong \lambda/\sqrt{2n\Delta n + (\lambda/2d)^2} \tag{1a}$$

$$\alpha_R \propto (m+1)^2 \lambda^2/d^3 \sqrt{\Delta n} \tag{1b}$$

where d is the antiguide-core width, Δn is the lateral refractive-index step, λ is the vacuum wavelength, and m is the (lateral) mode number. For typical structures ($d = 3\,\mu$m, $\Delta n = (2-3) \times 10^{-2}$) at $\lambda = 0.85\,\mu$m typical λ_1 and α_R values are 2 μm and 100 cm^{-1}, respectively. Since $\alpha_R \propto (m+1)^2$, the antiguide acts as a lateral-mode discriminator. For a proper mode to exist α_R has to be compensated for by gain in the antiguide core.[6] Single antiguides are currently used for mode control in CO_2 "waveguide" lasers.

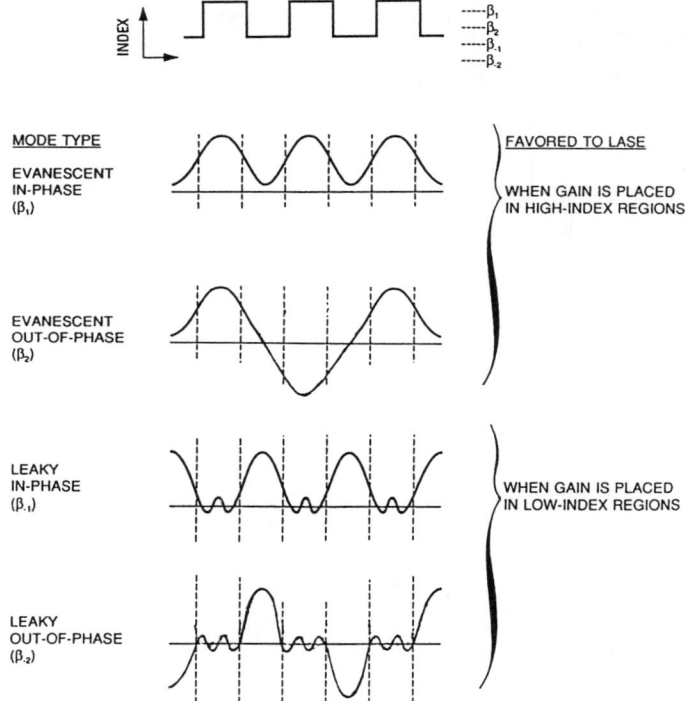

Fig. 3. Types of array modes for a periodic variation of the index of refraction.

A periodic variation of the refractive index (top of Fig. 3) represents arrays of both positive-index guides and negative-index guides. The supported array modes are evanescent-wave type when the fields are peaked in the high-index regions, and leaky-wave type when the fields are peaked in the low-index regions (Fig. 3). Solving for leaky array modes can only be done by using exact theory.[8] Eliseev, Nabiev and Popov[9] were the first to solve for both array mode types by using the Bloch-function method. Depending on preferential gain placement one or the other array-mode type is favored to lase. For antiguided-array modes to prevail gain is placed in the low-index regions,[10–12] with the high-index regions being transparent[10] or made lossy on purpose.[11,12]

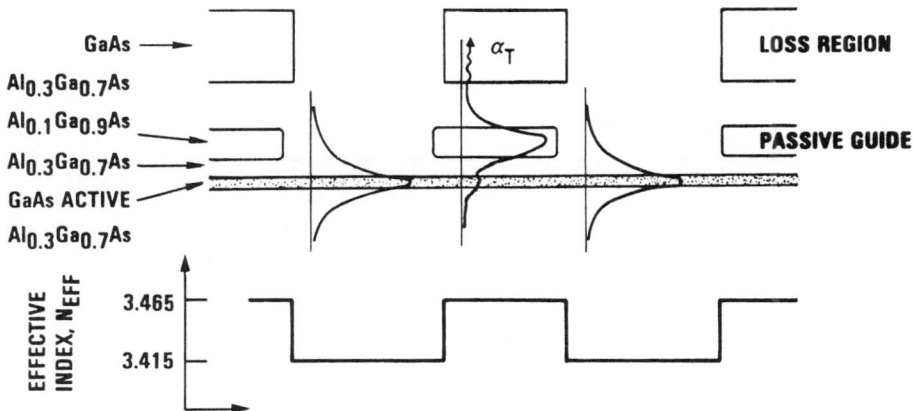

Fig. 4. Schematic representation of antiguided-array structure. Interelement low modal gain and optical losses suppress evanescent-wave array models.

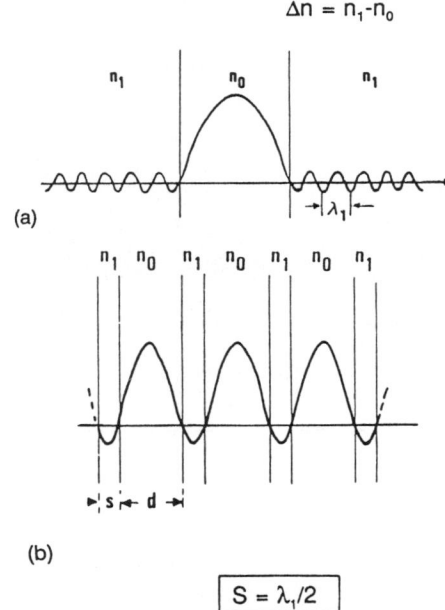

Fig. 5. Schematic representation of: (a) single antiguide (λ_1 is the leaky-wave "lateral" wavelength), (b) resonant array corresponding with the interelement spacing, s, corresponding to $\lambda_1/2$.

A practical way to suppress evanescent-wave array modes, and thus allow only leaky-mode oscillation, is shown in Fig. 4 for GaAs/AlGaAs structures. The interelement regions contain high-index passive-guide layers placed in close proximity ($0.1 - 0.2\,\mu\text{m}$) to the active layer. There the fundamental transverse mode is primarily confined to the passive guide layer. That is, between elements the "modal" gain is low. To further suppress oscillation of evanescent-wave modes an

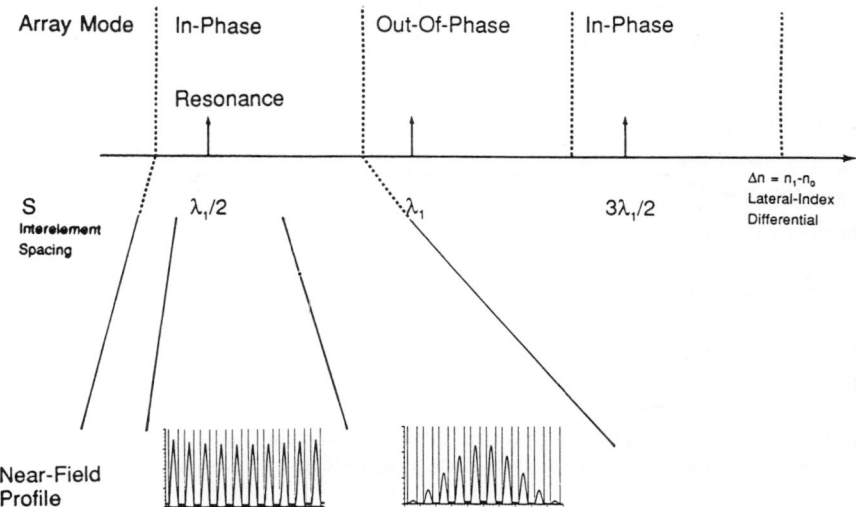

Fig. 6. The behavior of arrays of antiguides as the interelement region varies. The in-phase-mode near-field intensity profiles are shown near and far from resonance.

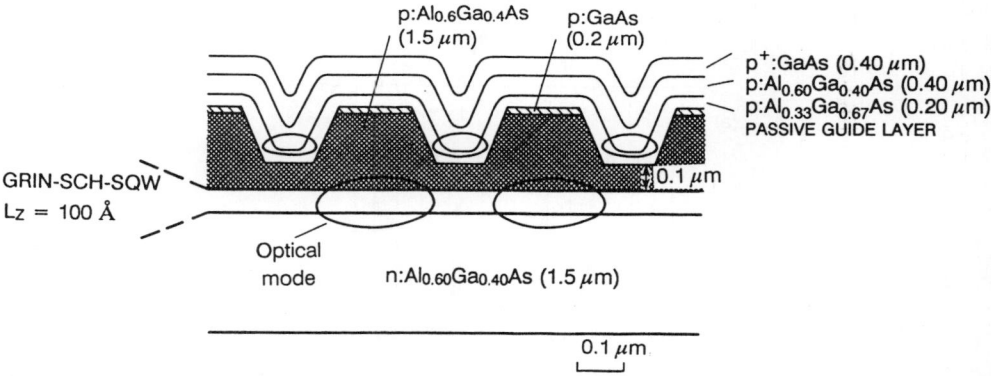

Fig. 7. Typical cross-section of ROW array.

optically absorbing material can be placed between elements.[8] A remarkable property of leaky-wave-coupling is the presence of a lateral resonance condition.[5,14] Then lateral radiation leakage antiguide (Fig. 5(a)) can be used for coupling multiple antiguides together. All elements resonantly couple in-phase or out-of-phase when the interelement spacings correspond to odd or even integral number of (lateral) half-wavelengths ($\lambda_1/2$), respectively (Fig. 5(b) and 6). The resonance condition is:

$$s = m\lambda_1/2, \quad m = \text{odd} \quad \text{resonant in-phase mode}$$
$$m = \text{even} \quad \text{resonant out-of-phase mode} \quad (2)$$

where s is the interelement spacing. Typical s values are 1 μm, as limited by photolitography. Then, for in-phase-mode resonance $\lambda_1 = 2$ μm. At resonance the interelement spacings behave as "half-wave" plates (i.e. Fabry–Perot resonators in the resonant condition). resonant devices are called resonant-optical-waveguide (ROW) arrays.[13] Fig. 6 shows what happens as s varies. In-phase and out-phase operational domains alternate. At and near resonance, due to parallel coupling, the near-field intensity profile is uniform, which in turn provides for uniform gain usage and maximum intermodal discrimination. Away from resonance the near field is cosine shaped due to poor overall coherence. Operating near resonance insured sole in-phase-mode operation. That is, ROW arrays act as excellent selectors of the in-phase array mode. A good analogy can be made to distributed feedback (DFB) lasers. While DFB lasers select a single longitudinal spatial mode, ROW arrays select a single lateral spatial mode.

The typical cross-section of a ROW array is shown in Fig. 7. A GRIN-SCH-SQW base structure is grown by MOCVD. Then etching and MOCVD regrowth are performed to create the high-index passive-guide regions between elements. Typical 20-element devices are fabricated with 3 µm-wide elements separated by 1 µm-wide regions. Best results are achieved for 1000 µm-long devices. To achieve resonance the Al concentration of the regrown passive guide layer is varied.

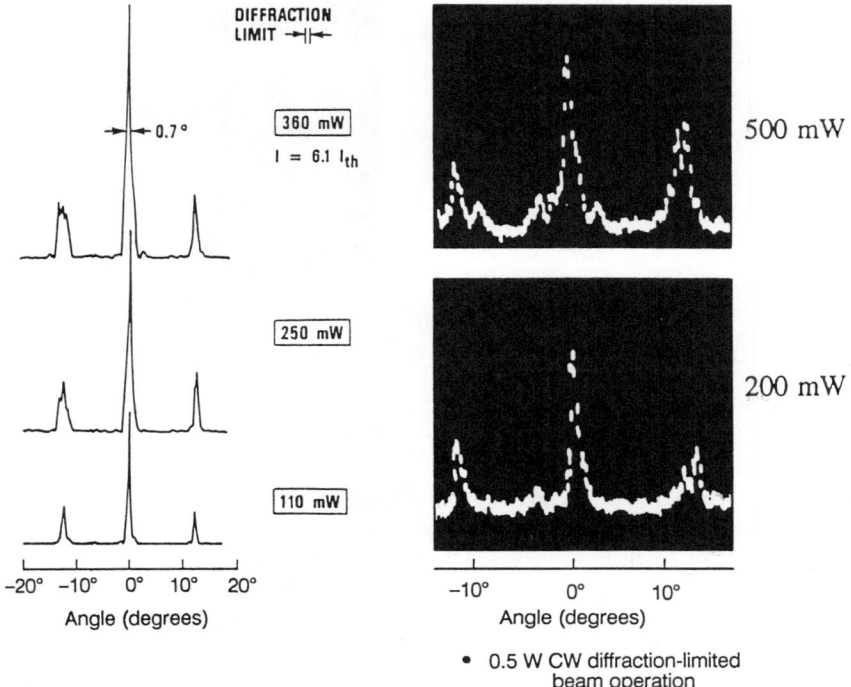

Fig. 8. CW far-field patterns of ROW arrays.

Best results to date are shown in Figs. 8–10. In CW operation a diffraction-limited beamwidth ($\Theta_{1/2} \sim 0.8°$) is obtained to 0.5 W, a world record. Light-current characteristics (Fig. 9) reveal high external differential quantum efficiency (42%), and high "wall plug" efficiency (22%). In pulsed operation the results are of world-record nature as well: 1.5 W of diffraction-limited power up to 10 × threshold[14] (Fig. 10).

For near-resonant devices spatial filters based on the Talbot effect need to be used.[15] Results are summarized in Fig. 11. For devices with Talbot filters[16] diffraction-limited operation occurs

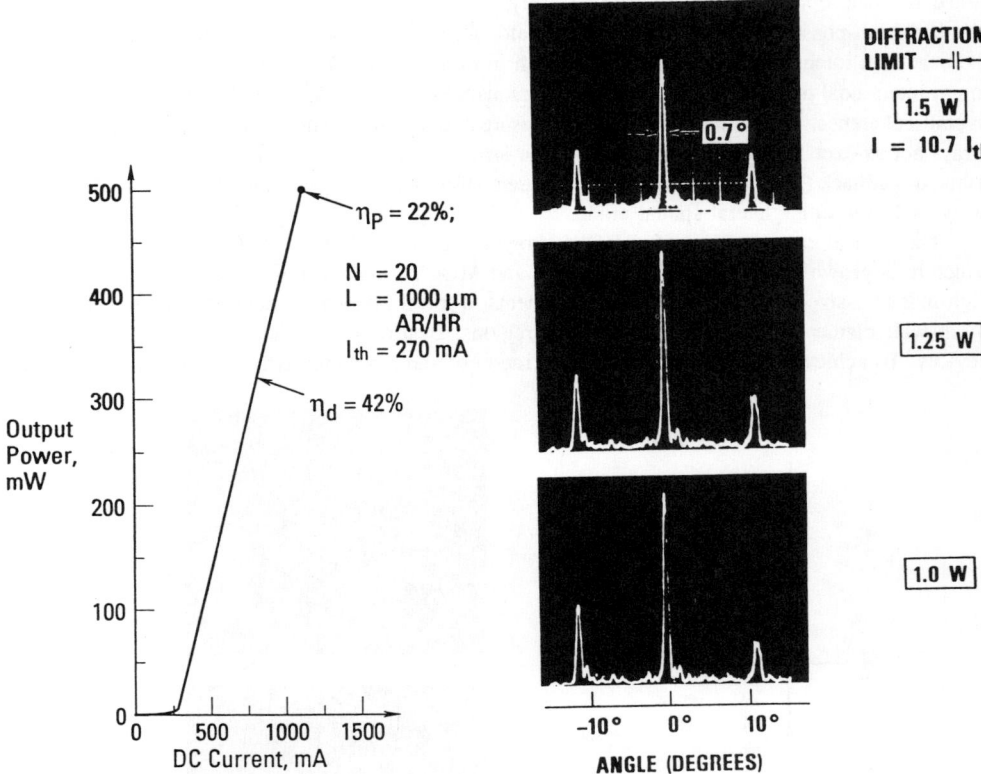

Figure 9. CW light-current characteristic of ROW array.

Figure 10. Pulsed far-field patterns of ROW arrays.

to 1.25 W, and beams 1.5 × diffraction limit are achieved at 2 W. For uniform devices[16] the beam is (2–3) × diffraction limit up to 5 W and 45 × threshold.

We typically use 20-element devices. For ≥ 40-element arrays it is hard to achieve good intermodal discrimination. Then one can further use resonant leaky-wave coupling to scale the coherent power (Fig. 12). As shown in Fig. 12, as long as interarray regions correspond to an odd number of $\lambda_{1/2}$ resonant long-range coupling is possible. Loss in the interarray regions suppresses nonresonant modes, and thus single-mode operation is achievable from large-aperture (> 100 μm) devices. Results of a preliminary experiment[17] are shown in Fig. 13 and 14. At 4-unit device was employed (Fig. 13). Without interelement gaps a 40-element array operated resonant out of phase: Fig. 14(a). With 21 μm-wide interelement gaps one obtains a pattern with the proper periodicity and beam quality expected for four coherently coupled 10-element arrays spaced 21 μm apart: Fig. 14(b). A fringe visibility close to 100% confirms locking across a 250 μm-wide aperture.

In conclusion, ROW arrays appear extremely promising as sources of watt-range coherent powers. Single-unit devices will soon provide 1 W CW diffraction limited. Beyond 1 W one requires scaling via resonant leaky-wave coupling. Preliminary results show coherence over a 250 μm-wide aperture, and thus the capability of achieving ≥ 4 W diffraction limited.

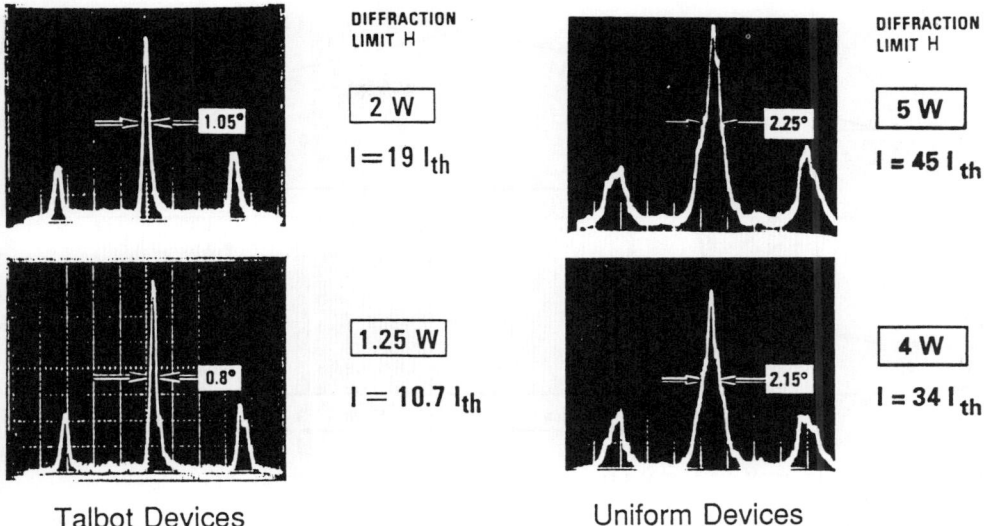

Fig. 11. Pulsed far-field patterns of near-resonant antiguided arrays with and without intracavity Talbot-type spatial filters.

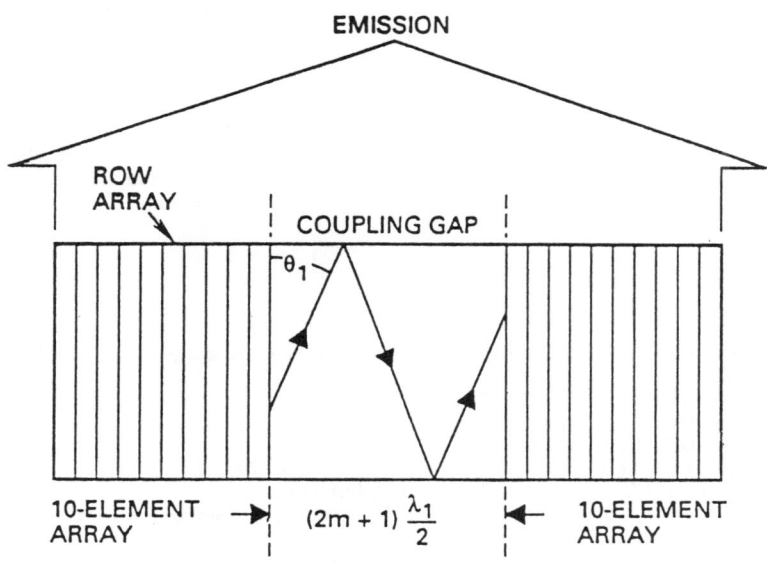

Fig. 12. Schematic representation of scaling via resonant leaky-wave coupling.

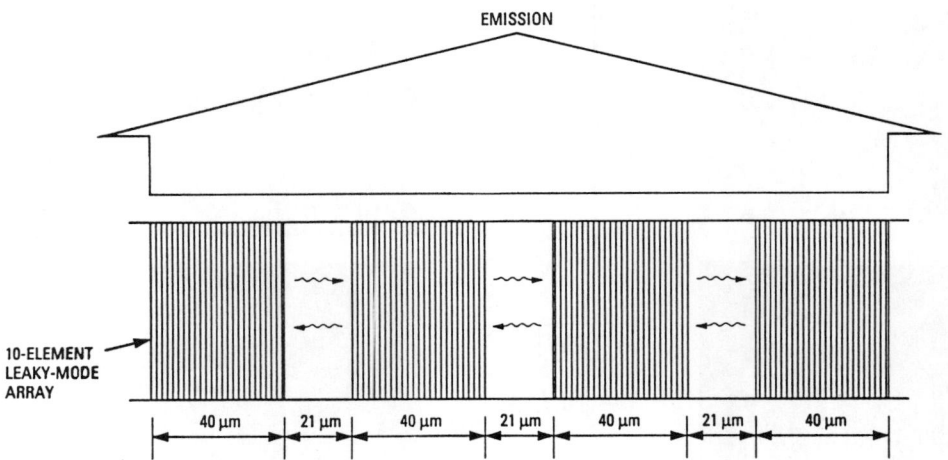

Fig. 13. Four-unit array coupled via resonant leaky-wave coupling.

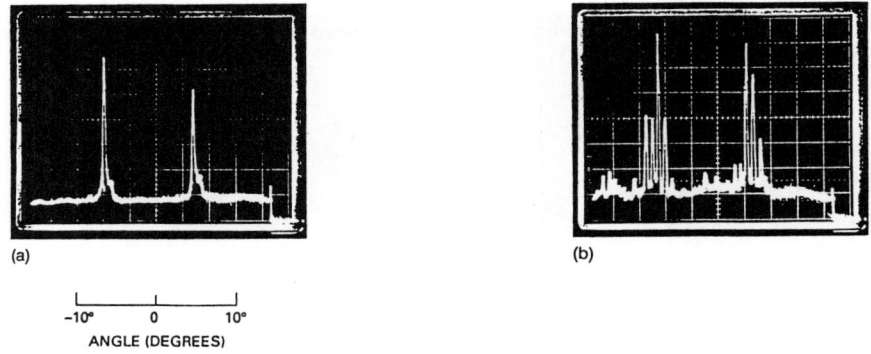

Fig. 14. (a) far-field pattern of 40-Element ROW array operating in the out-of-phase condition; (b) far-field pattern of four 10-Element ROW arrays coupled via 21 μm-wide gaps (see Fig. 13).

References

[1] S. Mukai, C.L. Lindsey, J. Katz, E. Kapon, Z. Rav-Noy, S. Margalit, and A. Yariv 1984 *Appl. Phys. Lett.* **45** 834
[2] D. Botez, T. Pham, and D. Tran 1987 *Electron. Lett.* **23** 416
[3] D. Welch, W. Striefer, and D. Scifres 1989 *Optics News* **3** 7
[4] W.J. Fader and G.E. Palma 1985 *Opt. Lett.* **10** 28
[5] D. Botez, L.J. Mawst, and G. Peterson 1988 *Electron. Lett.* **24** 1328
[6] R.W. Engelmann and D. Kerps 1980 *IEE Proc.* **127, Pt.I** 330
[7] V.V. Bezotosnyi, L.M. Dolginov, P.G. Eliseev, B.N. Sverdlov, E.G. Shevchenko, and G.V. Shepikina 1981 *Sov. J. Quantum. Electron.* **11** 1208
[8] D. Botez, L.J. Mawst, G.L. Peterson, and T.J. Roth 1990 *IEEE J. Quantum Electron* **QE-26** 482
[9] P.G. Eliseev, R.F. Nabiev, and Yu.M. Popov 1989 *J.Sov. Las. Res.* **10** 449
[10] D.E. Ackley and R.W.H. Engelmann 1981 *Apll. Phys. Lett.* **39** 27

[11] D. Botez, L.J. Mawst, P. Hayashida, G. Peterson, and T.J. Roth 1988 *Appl. Phys. Lett* **53** 464
[12] G.R. Hardley 1989 *Opt. Lett.* **14** 308
[13] D. Botez, L.J. Mawst, G. Peterson, and T.J. Roth 1989 *Appl. Phys. Lett.* **54** 2183
[14] L.J. Mawst, D. Botez, M. Jansen, T.J. Roth, and J. Rozenbergs 1991 *Electron. Lett.* **27** 369
[15] L.J. Mawst et.al 1989 *Electron. Lett.* **25** 365
[16] D. Botez, M. Jansen, L.J. Mawst, G. Peterson, and T.J. Roth, 1991 *Appl. Phys. Lett.* **58** 2070
[17] L.J. Mawst, D. Botez, M. Jansen, M. Sergant, G. Peterson, and T.J. Roth (in press)

Low threshold quantum well AlGaAs-heterolasers fabricated by low temperature liquid phase epitaxy

V.M. Andreev, A.B. Kazantsev, V.R. Larionov. V.D. Rumyantsev, and V.P. Khvostikov

A.F. Ioffe Physico-Technical Institute, Academy of Sciences of the USSR,
26 Polytekhnicheskaya st. 194021 Leningrad, USSR

Abstract. A decrease of AlGaAs layer crystallization temperature in LPE down to 550–400 °C allowed us to crystallize quantum wells as thin as 5–20 nm. Structures with a narrow-gap active layer of 15 nm thick were used to fabricate separate-confinement buried heterolasers. Low threshold currents $I_{th} = 2$ mA in a continuous regime (300 K) were measured for lasers with 100 μm long cavities. The maximum external quantum efficiency of radiation was 68% for a laser with $I_{th} = 2.3$ mA.

1. Introduction

Among semiconductor lasers, the best characteristics have been obtained for heterolasers with quantum wells (QW's) in the active region. At present AlGaAs-heterostructures for low threshold lasers are grown mainly by molecular beam epitaxy and organometallic chemical vapor deposition. These methods allow to prepare milliamper range stripe-geometry lasers and QW-heterolasers with a threshold current density lower than $j_{th} = 100$ A/cm^2.

This paper is concerned with low threshold AlGaAs-heterolasers prepared by low temperature liquid phase epitaxy (LT LPE).

This method allows to prepare AlGaAs-heterostructures with the active layer thickness 10–15 nm and to design on their basis diode lasers with $j_{th} = 120$ A/cm^2 (300 K). The advantage of these lasers is very low j_{th} for small (100–200 μm) cavities, which is important for reducing the absolute threshold current.

2. Low temperature LPE (technique and process)

For preparation of AlGaAs QW-heterostructures we use a technique developed earlier.[1] The main advantages of this technique (Fig. 1) over the conventional "sliding boat" method for low temperature LPE are as follows. A melt, before getting onto the substrate (1), is squeezed through a slot (2) by a piston (3), to provide an effective mechanical cleaning of the melt from an oxide film. This solves the problem of low-temperature LPE of AlGaAs-heterostructures, because it is very difficult to obtain an ideal substrate wetting at 400–500 °C. The mechanical cleaning can improve the wetting in spite of the presence of an aluminium oxide film on the melt surface. The melt substitution is performed by replacing one melt by another rather than by pulling the previous melt

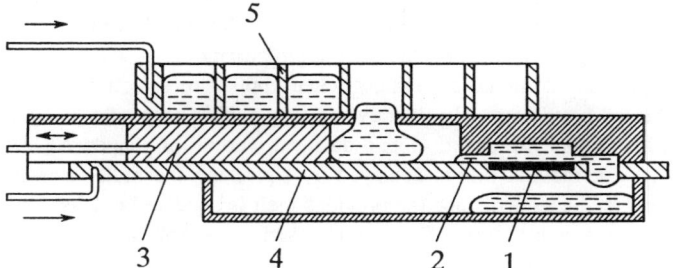

Fig. 1. "Piston" boat for LT LPE. (1) substrate; (2) input slot; (3) piston; (4) substrate holder; (5) container for melts.

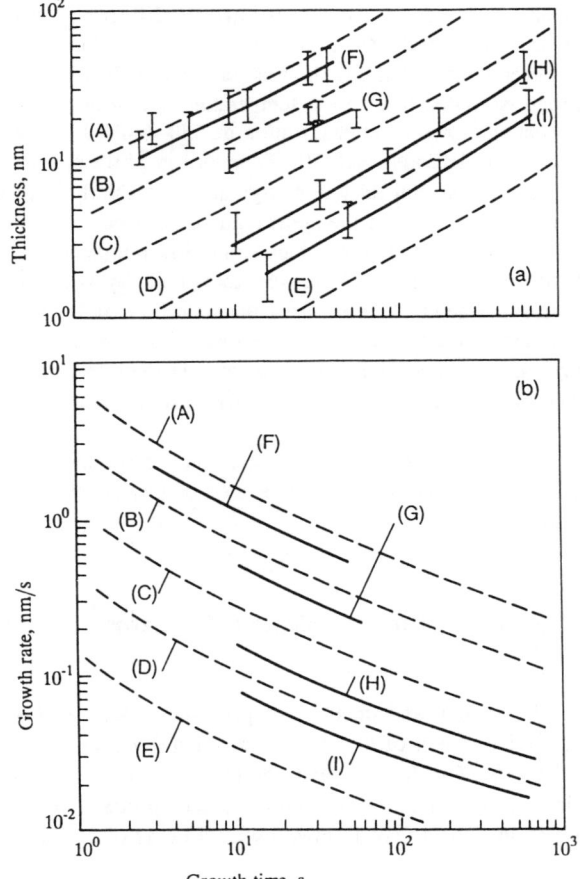

Fig. 2. Dependences of the layer thickness (*a*) and growth rate (*b*) of GaAs on the growth time for different crystallization temperatures (°C): (A) 600, (B) 550, (C) 500, (D,H) 450, (E,I) 400, (F) 560, (G) 530. The full curves (F–I) are experimental, all the dashed curves (A–E) are theoretical.

off the substrate and subsequent wetting. This reduces the number of wetting macrodefects. Crystallization can be carried out from a very thin melt layer (down to 100 μm) between the substrates. This provides a small growth rate and the structure planarity.

The technique has additional advantages when applied to selective LPE, which is used for preparation of buried heterolasers. In selective epitaxy, which necessitates cleaning the melt from the oxide film, it is rather difficult to obtain an ideal wetting of shaped AlGaAs-heterostructures in local areas. In our method this is achieved by squeezing the melt through a narrow slot in front of the substrate. Secondly, since the previous melt is squeezed out by the subsequent one, it provides more effective substitution of the melt, especially in hollows, when the layers are grown on a corrugated structure.

Fig. 2 shows the calculated dependences of GaAs layer thickness (Fig. 2(a)) and crystallization rate (b) on the growth time for the initial growth temperatures 600–400 °C. The calculations were made for a melt thickness of 0.1 cm, cooling rate 10^{-2} deg/s, supercooling of the melt by $\Delta T = 10$°C, and the value $D_{As} = 5 \cdot 10^{-6}$ cm^2/s for arsenic diffusion coefficient. The decrease in crystallization rate at the initial stage ($t < 10^2$ s) results from elimination of the supersaturation condition due to supercooling of the melt. The subsequent decrease in the growth rate is primarily due to the progressive decrease in the temperature and solubility of As in Ga. It can be seen from Fig. 2(b) that the rates of crystallization at 550–400 °C are of the same order of magnitude (10^{-1}–1 nm/s) as in MBE and MOCVD. Comparison of the experimental data with the calculations shows that the experimental thickness is greater than the theoretical values. This can be accounted for by greater supercooling of the melts than 10 °C accepted in the calculations. The theoretical and experimental values for the layer thickness at low temperatures ($T = 400-550$ °C) show that it is possible to fabricate AlGaAs heterostructures with layers as thin as 2–20 nm from a melt which is 10–20 °C supersaturated for the crystallization time of 10–30 seconds. The reproducibility is improved by decreasing the substrate size down to 2 cm^2 and the melt volume down to 0.2–0.3 cm^3. These results can be attributed to a decrease in the effect of non-uniformity of the melt supersaturation. The planarity of the layers is improved by decreasing the total thickness of the structure and buffer layers down to 1–2 μm.

So the method of low temperature LPE considered here is suitable for producing AlGaAs heterostructures with quantum wells (QW's) and fabrication of low-threshold separate-confinement double-heterostructure lasers.

3. Photoluminescent quality control of AlGaAs QW heterostructures grown by low-temperature LPE

Analysis of the spectral distribution and intensity of photoluminescence of AlGaAs heterostructures with a narrow gap active region of $d = 5 - 10$ nm is a good way to control the planarity and crystal perfection of QW's.[3] On the one hand, the theory gives us the thickness dependences of the energy gaps which can appear in the photoluminescece spectra. On the other hand, the efficiency of radiative recombination in QW's grown by LT LPE is of fundamental importance for laser operation.

Fig. 3 shows the photoluminescence spectra of a sample with $d \simeq 5$ nm at $T = 77$ K. The inset shows the AlAs distribution measured beginning from the sample surface by the method of x ray excited photoelectron emission. At low excitation density (Ar-laser, equivalent to 3 A/cm^2, curve A), the photoluminescence spectrum consists of a single band associated with the transitions from the first electron level to the first level of heavy holes (E_{1h}). The band for the light holes

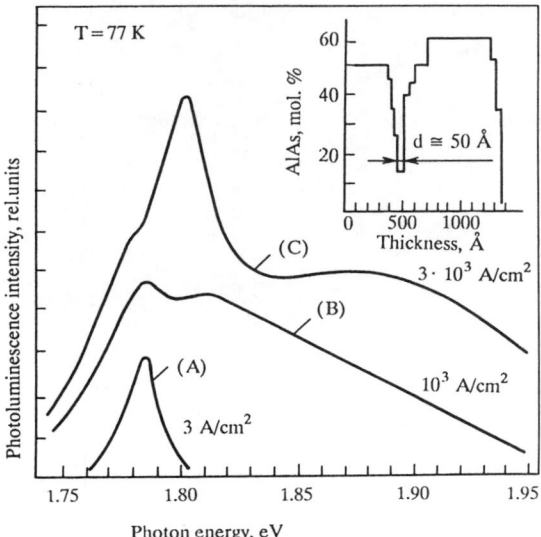

Fig. 3. AlAs distribution throughout the structure thickness and the photoluminescence spectra of a sample with $d \simeq$ 5 nm. The photoexcitation density values are taken as the equivalent densities of the electric current.

appeared at a high excitation density (N_2-laser, curve B) owing to a large energy gap between the luminescence bands representing transitions to levels of heavy and light holes. When the excitation density was equivalent to $3 \cdot 10^3$ A/cm² (curve C), there were both a high energy band (> 1.9 eV) for transitions from the second electron level and a stimulated emission band ($\simeq 1.8$ eV) in the spectrum. It must be noted that to achieve the filling of high-energy states at high excitation densities, it was necessary to ensure intensive longitudinal optical losses to block the stimulated emission, so that the QW was located at a distance of just $\simeq 0.1\,\mu$m from the absorbing GaAs substrate. Comparison of experimental half-widths for the low excitation luminescence bands with the theoretical dependence of the $E_{1h} - E_g$ value on d shows that the range of fluctuations of the QW thickness should not exceed ±1 monolayer for the samples grown by low-temperature LPE. On the other hand, the dependences of luminescence intensity on the excitation density at $T = 77$ and 300 K gave an estimation of the internal quantum efficiency of luminescence close to 100% in the range of excitation densities > 100 A/cm² at room temperature. So the geometrical and luminescent properties of LT LPE fabricated QW AlGaAs heterostructures do not limit the heterolaser operation at the expected threshold current densities of $\simeq 100$ A/cm².

4. Threshold current density (j_{th}) of AlGaAs heterolasers fabricated by LT LPE

We shall now consider our results on the characteristics of low-threshold lasers with a wide stripe contact.[4,5]

Fig. 4(a) shows a SEM micrograph of a cleaved section through an AlGaAs heterolaser. Fig. 4(b) gives a typical profile of AlAs distribution across infra-red laser structures. We used an n-type ⟨100⟩ GaAs substrate ($N_D \simeq 10^{18}$ cm³) to grow first n-type GaAs ($\simeq 0.5\,\mu$m) and n-type $Al_{0.7}Ga_{0.3}As$ ($\simeq 1.3\,\mu$m) layers doped with tellurium. These were followed by $Al_{0.3}Ga_{0.7}As$

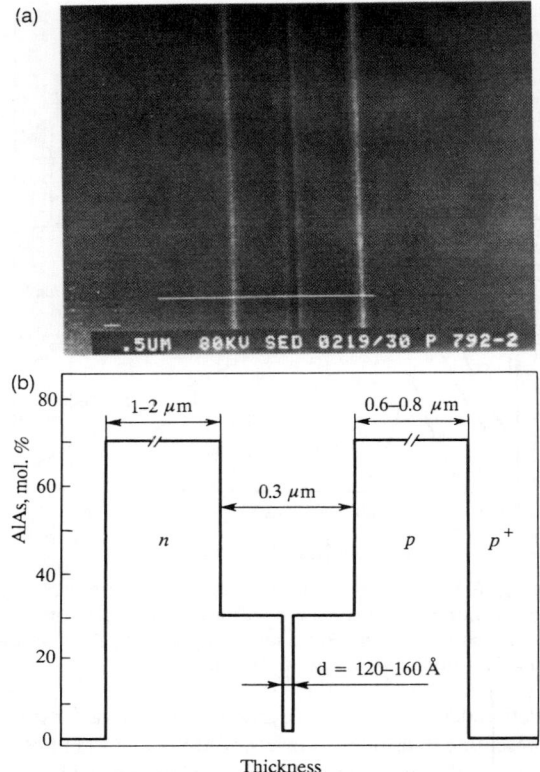

Fig. 4. Cleaved section (*a*) and profile of AlAs distribution across the thickness of an infra-red laser structure (*b*).

waveguide layers and an active region (with up to 5% AlAs in the solid phase for infra-red lasers) which were not doped. The upper layer of the wide-gap *p*-type $Al_{0.7}Ga_{0.3}As$ emitter was doped with magnesium and the p^+-type GaAs contact layer was doped with germanium. The active region crystallized at 545 °C with a thickness $d = 12 - 16$ nm. The last p^+-GaAs contact layer ($\simeq 0.2\,\mu m$) crystallized at 450–400 °C. We used metal contact Au-Ge/Au coatings with the substrate and Au-Zn/Au with the p^+-type GaAs layer. Wide stripe contact geometry lasers ($D = 50 - 100\,\mu m$) were fabricated. The measurements of j_{th} were carried out in the pulsed regime ($\tau_p = 1\,\mu s$, repetition rate $f = 1$ kHz).

In Fig. 5 the black dots represent the values of j_{th} ($T = 300$ K) for laser diodes, which were cleaved from several epitaxial wafers and had different resonator lengths (L). The minimum value of j_{th} was $j_{th} = 120\,A/cm^2$ at $L = 750\,\mu m$. Curve A in Fig. 5 connects the points corresponding to the best (lowest) values of j_{th} obtained by reducing the Fabry–Perot resonator length. Curve B gives the theoretical dependence of j_{th} on L for AlGaAs heterolasers with AlAs distribution approximately similar to Fig. 4(*b*) and $d = 10$ nm. It is seen that good agreement between the theoretical and experimental j_{th} values takes place for the samples with short resonators. As for the samples with long resonators, the j_{th} values are sufficiently low to fall within the range of the best results for the AlGaAs heterolasers fabricated by molecular-beam and vapor phase metalorganic epitaxial methods. However, these values are higher than those predicted by theory. In our case this fact is due to perceptible intracavity optical losses in the range $L > 500\,\mu m$, because

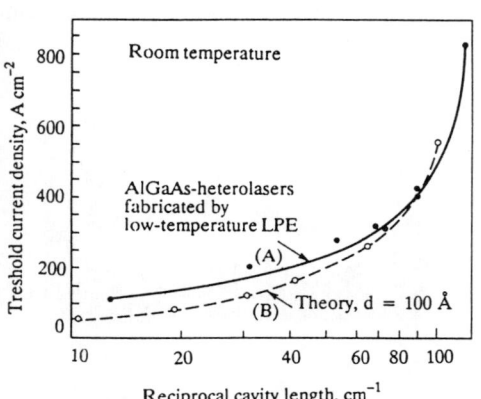

Figure 5. The best j_{th} values for LT LPE fabricated heterolasers versus reciprocal cavity length L^{-1} and the theoretical dependence $j_{th} = f(L^{-1})$ for an AlGaAs heterolaser with $d = 10$ nm.

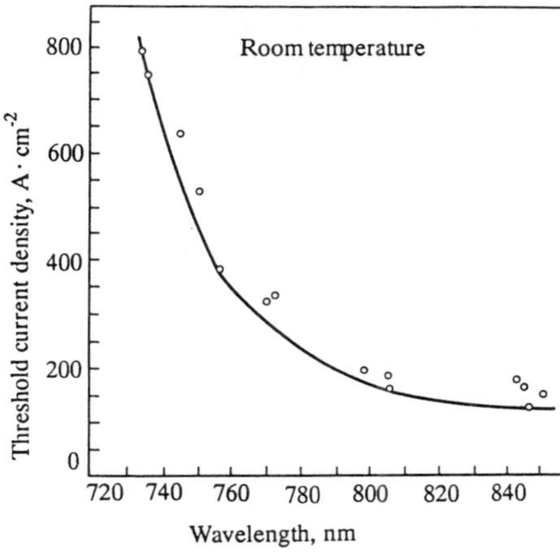

Figure 6. Dependence of j_{th} at room temperature on the emission wavelength plotted for AlGaAs heterolasers with different AlAs contents in the active QW region.

there is a partial penetration of the waveguide mode into the absorbing p^+-type GaAs layer across the wide-gap p-type emitter of 0.5–0.6 μm thick.

The j_{th} values versus the wavelength λ of emitted light ($T = 300$ K; $L = 600 - 900\,\mu m$) are plotted in Fig. 6 for lasers with varying AlAs content in the active QW region (from 3% to 22% at λ of 850 to 730 nm, respectively). It is seen that AlGaAs heterolasers fabricated by LT LPE have excellent threshold current characteristics within the practically important spectral range λ = 750 – 850 nm.

5. Milliampere-range CW long lifetime AlGaAs heterolasers

It is well known that in order to reduce the absolute value of threshold current (I_{th}), it is necessary to narrow down the stripe width of heterolasers and to cleave moderately short Fabry–Perot resonators ($L < 300\,\mu m$). Obviously (see Fig. 5), the LT LPE grown AlGaAs heterolaser structures are very promising for this purpose.

At first we shall consider the results concerning ridge-guide laser diodes.[6]

A ridge-guide laser was produced using a self-aligned etching process and anodic oxidation of passive areas, so that the current channel width was $D = 4.5\,\mu m$ (Fig. 7). After that all-round ohmic contacts were deposited on both sides of the wafer. The laser samples with and without the reflecting coatings on the cleaved resonator facets were studied. The coatings were formed by depositing three SiO_2-Si layer pairs with the total reflection coefficient $R \simeq 0.95$ in the spectral range $\lambda = 0.8 - 0.85\,\mu m$.

Fig. 7 shows also the dependences of $I_{th} = f(L)$ obtained for laser diodes with different reflection coefficients of the Fabry–Perot resonator mirrors. All samples were prepared on the basis of structures with a threshold current density $j_{th} = 170$ A/cm^2 at $L = 700\,\mu m$ and $D = 50\,\mu m$ in control diodes. In narrow ridge-guide lasers ($D = 4.5\,\mu m$) the minimum I_{th} was 5.5 mA for samples without reflecting coatings on the rear and front facets (Fig. 7, curve A; $R_{rear} = R_{front} \simeq$

0.3; $L = 70\,\mu\text{m}$; $\lambda = 825$ nm; $T = 300$ K). The presence of reflecting coatings allowed to reduce the I_{th} values by reducing the output optical losses of the Fabry–Perot resonator. So, for lasers with high-reflecting rear mirrors the minimum I_{th} was 3.4 mA (curve B), whereas for a laser with both rear and front coated mirrors this value was as low as 2 mA (curve C).

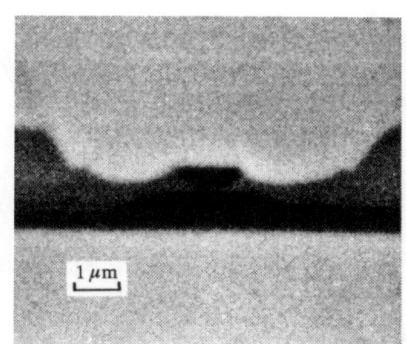

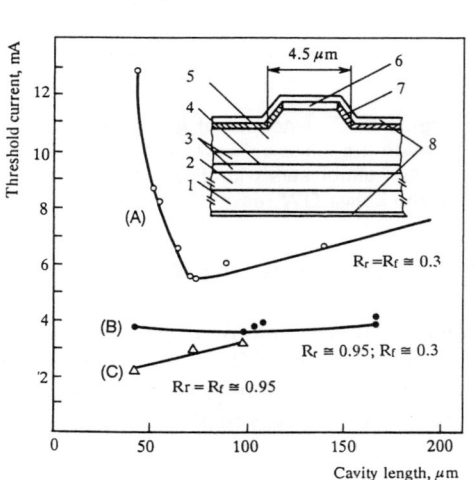

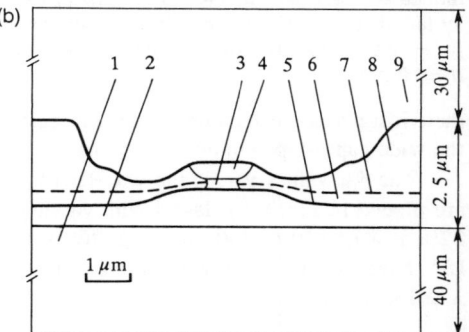

Figure 7. Schematic diagram of a ridge-guide laser: (1) n-GaAs substrate and buffer layer; (2) n-Al$_{0.7}$Ga$_{0.3}$As; (3) $n°$-Al$_{0.3}$Ga$_{0.7}$As; (4) Quantum well $n°$-Al$_x$Ga$_{1-x}$As; (5) p-Al$_{0.7}$Ga$_{0.3}$As; (6) p^+-GaAs; (7) Anodic oxide; (8) Ohmic contacts. Dependences of $I_{th} = f(L)$ obtained for laser diodes with different reflection coefficients for rear (R_r) and front (R_f) facets.

Figure 8. SEM micrograph (a) and its schematic diagram (b) of a "twice-buried" heterolaser: (1) n-GaAs substrate and buffer layer; (2) n-Al$_{0.7}$Ga$_{0.3}$As; (3) waveguide region including QW; (4) p-Al$_{0.7}$Ga$_{0.3}$As; (5) boundary of etching in non-saturated Ga+Al melt during the second LPE process; (6) p-Al$_{0.35}$Ga$_{0.65}$As; (7) line isolating the p-n junction; (8) n-Al$_{0.35}$Ga$_{0.65}$As; (9) thick p^+-GaAs cap layer grown during the third LPE process.

Further reduction of I_{th} was achieved by using a specially developed technique for making "twice-buried" stripe geometry lasers. The basic ideas underlying the laser design were as follows: to narrow down considerably the width of the current channel; to confine the optical mode in the lateral directions; to compensate for the strain tensions in the active QW in order to solve the laser lifetime problem. A SEM micrograph and its schematic diagram for a "twice-buried" heterolaser is shown in Fig. 8(a), (b). The structure was fabricated in three LPE processes. At first, a planar laser structure similar to that shown in Fig. 4 was grown by low-temperature LPE. After masking with SiO$_2$ stripes of 4–6μm width, the wafer was exposed to another low-temperature LPE treatment (but the first one formed the buried structure). Here, the areas without protection were etched by non-saturated Ga+Al melt down to a level shown as line 5 in Fig. 8(b). Owing to the selectivity of the etching in the lateral directions, waveguide stripes of about 1 μm width were left to form the current channels. After removing the etching melt, we made epitaxially grown p-$Al_{0.35}Ga_{0.65}As$ (6) and n-$Al_{0.35}Ga_{0.65}As$ (8) layers. The reverse biased p-n junction (7) formed in this process was used to isolate the passive areas of the heterolaser. Finally, after removing the SiO_2 mask, a thick ($\simeq 30 \mu m$) all-round p^+-GaAs layer was grown using a third LPE process. This process may be considered as the second one to bury the active QW and the waveguide region.

A standard procedure was used to thin the substrate down to 60–80 μm of the total wafer thickness and to deposit the ohmic contacts. It is important that some bending of the wafers should be observed after the thinning of samples with a small p^+-GaAs cap thickness. The presence of the bending means that the QW active layer, which is situated in the upper part of the wafer, is obviously stressed. At the same time, no bending took place in heterostructures with the cap layer as thick as 30 μm described above. This means that the presence of a thick p^+-GaAs cap allows to compensate for the tensile stress of the QW layer.

The output optical power versus current characteristics for the "twice-buried" lasers under pulse operation ($T = 300$ K) are shown in Fig. 9. These lasers had no reflecting coatings on the facets. It is seen that the lasers with very low I_{th} values within the range of several milliamperes (for example, $I_{th} = 2$ mA at $L = 100 \mu m$ and $I_{th} = 3.9$ mA at $L = 620 \mu m$; $\lambda = 815 - 820$ nm) are characterized by high external differential efficiency of radiation (per two facets) $\eta_e = 54 - 68\%$. These results are in a file of the best reported data on AlGaAs heterolasers without reflecting coatings. Application of coatings will allow to fabricate the submilliampere heterolasers.

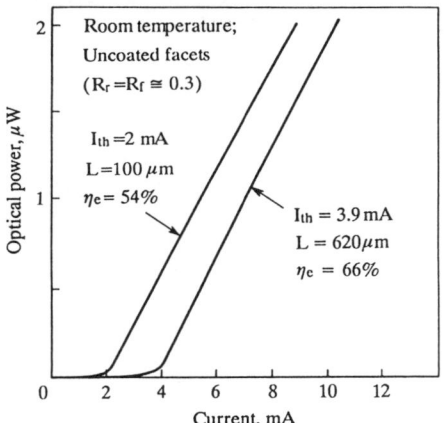

Fig. 9. Output optical power versus the diode current for "twice-buried" AlGaAs heterolasers fabricated by LT LPE.

Finally, we have estimated the lifetime of "twice-buried" lasers testing the devices under continuous wave operation at a higher temperature of the p-n junction (55–60 °C) and the output optical power 2 mW. For lasers with $I_{th} = 6 - 8$ mA the estimated lifetime was over 20,000 hours at room temperature.

6. Summary

We have successfully developed a LT LPE technique to fabricate high quality AlGaAs quantum well heterolaser structures and "twice-buried" stripe geometry heterolasers. Very low values of threshold current density and absolute threshold current (120 A/cm^2 and 2 mA, respectively) have been obtained in the lasers at high output efficiency with an estimated lifetime of over 20,000 hours. LT LPE seems to be a very promising and simple technique to fabricate perspective devices based on AlGaAs heterostructures with ultra-thin layers.

Acknowledgments

The authors are grateful to Zh.I. Alferov for encouragement and useful discussions. The authors also wish to thank V.B. Khalfin for making the theoretical j_{th} values available to us.

References

[1] Zh.I. Alferov, V.M. Andreev, S.G. Konnikov, V.R. Larionov, and B.W. Pushny 1976 *Kristall und Technik* **11** 1013
[2] Zh.I. Alferov, V.M. Andreev, A.A. Vodnev, S.G. Konnikov, V.R. Larionov, K.Yu. Pogrebitskii, V.D. Rumyantsev, and V.P. Khvostikov 1986 *Sov. Techn. Phys. Lett.* **12** 450
[3] V.M. Andreev, A.A. Vodnev, A.M. Mintairov, V.D. Rumyantsev, and V.P. Khvostikov 1987 *Sov. Phys. Semicond.* **21** 736
[4] Zh.I. Alferov, V.M. Andreev, V.Yu. Aksenov, V.R. Larionov, V.D. Rumyantsev, and V.P. Khvostikov 1988 *Sov. Phys. Semicond.* **22** 1123
[5] V.M. Andreev, V.Yu. Aksenov, A.B. Kasantsev, T.A. Prutskikh, V.D. Rumyantsev, E.M. Tanklevskaya, and V.P. Khvostikov 1990 *Sov. Phys. Semicond.* **24** 1096
[6] Zh.I. Alferov, V.M. Andreev, V.Yu. Aksenov, T.N. Nalet, Nguen Tkhan' Fyong, V.D. Rumyantsev, and V.P. Khvostikov 1988 *Sov. Tech.- Phys. Lett.* **14** 893

Estimation of output power from semiconductor laser limited by optical nonlinearity

P.G. Eliseev and R.F. Nabiev

P.N. Lebedev Physical Institute, Academy of Sciences of the USSR,
Leninsky pr. 53, 117924 Moscow, USSR

Abstract. Nonlinear focusing in semiconductor active medium of the laser is considered as an important cause of deterioration of properties and spatial stability of the laser emission beam. Self-focusing threshold power is estimated for different active region configurations, namely, for bulk, wave-guide and non-guide geometries. Output power just below the self-focusing threshold is assumed as a high-quality beam power limit, caused by optical nonlinearity in the semiconductor laser. It is shown, in particular, that in some cases non-guide configuration provides significantly higher power limit than a more traditional wave-guide configuration.

1. Introduction

Semiconductor active media of lasers have large optical nonlinearity, and self-focusing (SF) is one of nonlinear effects which may affect beam quality of the laser emission (see, e.g.[1]). The SF effect distorts the wavefront and deforms the spatial profile of the beam. In some cases SF leads to formation of optical "filaments" or to pulsation of the emission power. This outcome of nonlinear optical processes is unacceptable in the lasers designed for high-power operation, if high-quality (single-mode) beam is desirable.

Here we consider output power limitation in semiconductor lasers that is dependent on the geometry of the active region. The goal is to obtain criteria for designing high-power devices with stable and undistorted beam profile.

2. On the self-focusing mechanism

A simplified description of the SF effect is based on the models of linearized refractive index n, dependent on the intensity I:

$$n(I) = n_0 + n_2 I, \qquad (1)$$

where n_0 and n_2 are regarded as constants. When n_2 is positive, the beam becomes unstable at a certain sufficiently high intensity and a self-confined optical filament may appear.

G.H.B. Thompson[2] presumed to observe SF occurrence in stripe GaAlAs/GaAs lasers; the behavior of these lasers involving probable SF effect is described also by P.A. Kirkby et al.[3] Some experiments to indentify the SF in broad-area GaAlAs/GaAs lasers were performed by H. Bachert et al.[4]

To understand the nature of the nonlinear coefficient n_2 let us consider the influence of free carriers on the refractive index. Due to saturation of interband transition and to shift of plasma resonance, the regular contribution of the carriers at the operating wavelength of the laser is negative,

$$\sigma \equiv \frac{dn}{dN} < 0,$$

where N is the carrier concentration. σ is a useful quantity for further expressions. If the incident light produces photoelectric absorption in the solid, the local value of N increases and the refractive index may be reduced. This is the case of self-defocusing. But if the solid has inverse population, the resonantly interacting light may produce a decrease of the local value N instead of increase due to intensification of the stimulated emission. Therefore, the intensity increase in the latter case results in a decrease of the effective carrier lifetime τ, leading to a decrease of N and providing positive contribution to the refractive index. For n_2 one may write[3]

$$n_2 = -\sigma g \tau / h\nu, \tag{2}$$

where g is the gain coefficient, $h\nu$ the photon energy. This expression is valid for optical amplifier whereas for the laser oscillator a self-consistent state of both the electron and the photon systems must be taken in consideration.

In addition to electronic contribution to the optical nonlinearity thermal effects may be also considered. In GaAs and similar semiconductors local heating is accompanied by local increase of the refractive index. Thus, this effect may produce SF occurrence, and it may be treated in terms of nonlinear coefficient n_2, since heating is caused by dissipative absorption of light. The process is significantly slower than the electronic one, and in a semiconductor laser it is combined with thermal lens effect induced by an inhomogeneous dissipation of the unconsumed pump power.

A general description of SF phenomena in semiconductor laser medium is very complicated. We shall limit our discussion to a simple model with constant n_2, which allows to obtain useful formulae for the estimation of the SF threshold power which will be considered in further Sections as an upper limit of output power of high-quality-beam laser, if this limit is less than that caused by optical self-damage of the material.

These power limit estimations are carried out in the hope to find improvements leading to higher power of the semiconductor lasers. At present there are no semiconductor lasers as powerful as some solid-state lasers (e.g. Nd:YAG-lasers) with high-quality single-mode laser beam; owing to this the diode lasers, having very high efficiency, are used as sources of optical pumping for solid-state lasers rather than as direct sources of the required laser beam.

3. One-dimensional SF in plane waveguide geometry

Earlier it was found that one-dimensional SF-problem has a solution with the transverse beam profile

$$u(y) = u_0 / \cosh(ay), \tag{3}$$

where y is the transverse coordinate, $u(y)$ the profile function for the field, u_0 and a are coupled parameters, so that the quantity u_0/a remains constant as the beam propagates along the axis z.

The solution was given by R.Y. Chiao et al.[5] and used for description of the SF effect in plane waveguide geometry of the semiconductor laser.[4,6] The power threshold of SF is estimated by

$$P^{1D} \approx 0.18\lambda^2 d_0 / n_0 n_2 w_0,$$

where d_0 is the thickness of the beam (along the normal to the waveguide plane), w_0 is the initial value of beam width. In the GaAs active medium P is about 10 mW if $\lambda = 850$ nm, $d_0 = 10^{-4}$ cm, $w_0 = 10^{-3}$ cm, $n_0 = 3 \cdot 10^{-9}$ cm^2/W, $n = 3.6$. This threshold is reasonably close to that of the typical kink occurrence in the first-generation gain-guided stripe laser on the base of GaAs/GaAlAs heterostructure.

4. Non-guide geometry

In an infinite nonlinear medium the SF threshold may be calculated using Gaussian beam approximation, which is not a general solution, but is acceptable for the estimation[7]

$$P^{2D} \approx \lambda^2 / 4\pi n_0 n_2. \qquad (4)$$

At the same quantitative parameters of the medium and wavelength it gives some 50 mW independent of the initial size of the beam profile.

There is an opportunity to improve the situation and to enhance the SF threshold if one uses the sectioned active medium with linear gaps between active nonlinear sections. In this case the diffraction in the gaps may introduce beam spreading which can compensate the beam narrowing due to SF effect. The concept of SF occurrence in the sectioned non-guided geometry is given in Ref. 8.

If one period of the sectioned medium contains a gap section with the length l and constant refractive index n and an active section with the length L and refractive index $n = n_0 + n_2 I$, the modified SF threshold may be expressed as[8]

$$P^* = P^{2D}(1 + n_0 L/nl)^2 [1 + Z_1^2(1 + 1/L)^2 + 2l/L] \qquad (5)$$

where $Z_1 = \lambda L / \pi n w_0^2$. For example if $Z_1 \ll 1$ and $L \gg 1$ one may find

$$P^*/P^{2D} \approx (n_0 L/nl)^2. \qquad (6)$$

Therefore, the SF threshold may be greatly increased. When $Z_1 \gg 1$ and $L \gg 1$ the improvement factor will be

$$P^*/P^{2D} \approx (L^2 n_0 \lambda / 2\pi n^2 l w_0^2)^2 \qquad (7)$$

even in steeper rise as compared to the ratio L/l.

Therefore, tailoring of active region of the laser with non-guide geometry may allow to increase significantly the power range before the SF occurrence, therefore, with no beam disturbances caused by the optical nonlinearity.

For detailed picture of SF occurrence additional considerations have to be regarded. They are as follows: (i) a single-pass SF occurrence in one section of the active region; (ii) small-scale SF occurrence concerned with self-imaging effect in free diffraction process in the linear gaps.

The estimation obtained in Ref. 8 shows that in particular cases a modified SF threshold may be as high as several kilowatts instead of 50 mW in infinite medium. Consider for example the configuration of the "radiating mirror" laser of relatively thin active layer placed at one mirror of external cavity. There is only a single period of the "sectioned" active region, but the linear gap has to be overpassed twice by the beam. This type of geometry may be used in the lasers with electron-beam or optical pumping. If $l = 0.1$ mm, $w_0 = 1$ mm and $L = 1$ cm (other parameters as above) one may find $P^* \approx 6.6$ kW, significantly higher than P^{2D} or P^{1D}. The estimations also give about 580 kW for a small-scale SF threshold and 120 kW for a single-pass SF threshold, therefore, the latter two are not so important as the power limits in comparison to the "large-scale" SF threshold P^*.

The concept of non-guide diode laser was reported in Ref. 9 with an example of geometry of the active region tilted to the cavity axis. This geometry allows the beam to propagate in the linear transport regions in the cavity with free diffraction. Therefore, this is also a particular case of the laser configuration with the enhanced SF threshold.

Calculations of a vertical-cavity SF lasers[10] are made also for an other case of geometry, improved by introducing the linear gap into the cavity. Typical value P^* is about 2 W for $l = 10^{-4}$ cm, $w_0 = 2 \cdot 10^{-4}$ cm, $L = 5 \cdot 10^{-4}$ cm; this threshold is probably higher than the self-damage limitation. Therefore, in this case one does not expect any SF occurrence up to the self-damage power threshold.

5. Conclusion

The SF effect is one of important nonlinear mechanisms responsible for power limitation in high-quality single-mode semiconductor lasers. It produces instabilities and profile deformations of the beam. The estimated threshold of SF appears to be relatively low in ordinary waveguide-geometry semiconductor lasers, and, also, in the non-guide configuration of the infinite active medium. The latter is the case when the cavity is completely filled by the nonlinear medium. The situation may be improved if the configuration includes linear gaps in the cavity alternatively with active sections along the cavity axis. In these cases the SF threshold may be significantly increased.

References

[1] A.P. Bogatov and P.G. Eliseev 1985 *Kvant. Elektron.* **12** 465
[2] G.H.B. Thompson 1972 *Optoelectronics* **4** 257
[3] P.A. Kirkby, A.R. Goodwin, G.H.B. Thompson, and P.R. Selway 1977 *IEEE J. Quantum Electron.* **13** 705
[4] H. Bachert, A.P. Bogatov, P.G. Eliseev 1978 *Sov. J. Quantum Electron.* **8** 346
[5] R.Y. Chiao, E. Garmire, and C.H. Townes 1964 *Phys. Rev. Lett.* **13** 479
[6] P.G. Eliseev 1983 *Introduction into Physics of Injection Lasers.* "Nauka", Moscow 103
[7] Y.R. Chen 1984 *The principles of nonlinear optics* J.Wiley and Sons, N.Y. 303
[8] P.G. Eliseev and R.F. Nabiev (to be published)
[9] A.P. Bogatov, P.G. Eliseev *et al.* 1979 *Kvant. Elektron.* **6** 2639
[10] H. Soda, K. Iga, C. Kitahara, and Y. Suematsu 1979 *Jpn. J. Appl. Phys.* **18** 2329

High-power grating tuned semiconductor diode lasers and single-frequency diode-pumped Nd:YAG microcavity lasers

P. Gavrilovič, S. Singh, V.B. Smirnitskii,[a)] J. Bisberg, and M. O'Neil

Microelectronics Laboratory, Polaroid Corp., 21 Osborn st., Cambridge, MA 02139, USA

Abstract. Data are presented on 100-μm aperture gain-guided single-quantum well laser diodes in an external grating cavity. A maximum power of 500 mW is coupled out of the cavity with an efficiency of 0.51 W/A. The laser emission has a linewidth of ≤ 1 Å and is tunable from 7950 to 8450 Å for the specific laser diodes used in this study. The output beam is collimated in the direction perpendicular to the p-n junction, and exhibits a divergence of 0.4 degrees parallel to the p-n junction. A single-frequency Nd:YAG microchip laser fabricated from flux-grown YAG is demonstrated. The microchip laser is pumped with the output of the external-cavity laser. The maximum power is 45 mW limited by the available pump power. The single-frequency output is temperature-tunable over a range of 6 Å.

1. Introduction

At present, high-power quantum well laser diodes in the AlGaAs/GaAs material system are widely available and are being used in an increasing number of applications. In this paper we will focus on two specific applications of such lasers. First, we describe a tunable external-resonator laser that uses a broad-stripe high-power laser diode as the active element. In the second part of the paper, we describe a new development in laser diode-pumped solid-state lasers, single-frequency Nd:YAG microcavity lasers. The microcavity laser is end-pumped with the focused output of the grating-tuned laser.

2. High-power grating-tuned laser diode

The use of an external dispersive cavity is a well-established means for narrowing and tuning the spectral output of semiconductor injection lasers.[1-8] Past work has been concerned primarily with obtaining narrow linewidth and a wide tuning range, and has employed low-power diodes. Rossi and coworkers[1] operated AlGaAs/GaAs single-heterojunction diodes up to an output power of 3.5 W in an external grating cavity, but the diodes were driven with low duty-cycle current pulses (duty cycle $\approx 10^{-4}$), and thus the average power was low. More recently, in experiments using quantum well (QW) laser diodes, a continuous-wave (CW) output power of 70 mW was obtained from a 100-μm wide multistripe device.[7] In the present paper we demonstrate a broad-band, high-power, grating-tuned AlGaAs/GaAs QW laser with a maximum CW power output of

550 mW and a tuning range of 500 Å. The tuning is continuous with a resolution that is determined by the mechanical resolution of the grating mount and the lasing linewidth. The output is a collimated beam with cross-sectional dimensions of several millimeters. This beam size is convenient for shaping with commonly available optical components, which is an important consideration in applications of the laser.

The laser diodes used in this work are single-quantum well separate confinement devices that are grown by low-pressure metalorganic chemical vapor deposition. The GaAs QW, 100–130 Å thick, is centered in an $Al_xGa_{1-x}As$ waveguide that is confined by higher-gap p- and n-doped $Al_yGa_{1-y}As$ cladding layers. A 100-μm wide active region stripe is defined by a 100-keV H^+ implant. The front and rear facets are coated with antireflective (AR) and high-reflectivity (HR) coatings having reflectivities of 0.02 and 0.95, respectively. The front-facet slope efficiency after coating is 1.0–1.1 W/A for the standard cavity length of 500 μm that is employed in theis work. The diode is mounted on a copper heat sink which is attached to a thermoelectrically cooled mount that controls the heat sink temperature to within $\pm$ 0.05 °C.

The external cavity is similar to designs that have been employed previously.[1-8] The cavity consists of a collimating objective lens and a Littrow-mounted grating in a 2-axis gimbal mount. The cavity length is 150 mm. The 0.65 N.A., AR-coated collimating lens has a 6.5 mm focal length, an 8-mm physical aperture, and transmits 95% of the front-facet emission. The collimated beam illuminates a grating that diffracts in first order. Operation in first order only is desirable because of the high feedback efficiency and output efficiency, as discussed below. The grating is oriented such that the spectrum is dispersed in a plane perpendicular to the diode's epitaxial layers (the grating lines are parallel to the active region), and thus the active region acts as the entrance slit of a monochromator in coupling the feedback into the diode.[9] The alternative mounting for the grating, with dispersion parallel to the active region, is not effective when used with side-stripe diodes, although it can be used with narrow-stripe devices.[5]

The diode emission is coupled out of the resonator through the zeroth-order (undiffracted) reflection from the grating. The advantage of this scheme is that no additional coupling elements, such as beamsplitters, are inserted into the cavity, thus maximizing the output efficiency and simplifying alignment. Furthermore, the present scheme is preferable to a design that uses a second lens to collect light from the diode facet opposite to the grating,[2,4,5] since that requires a more complicated diode package. Such a package requires accurate matching of diode chip and heat sink lengths and degrades the laser heat sinking, which is of great importance for reliable operation at high powers.

The efficiency of optical feedback into to the diode from the external cavity, η_f, can be written in terms of the grating reflectivity in first order, R_{g1}, the lens transmission, T_L, and the coupling efficiency β, as follows: $\eta_f = (R_{g1})(T_L^2)(\beta)$. In order to maximize feedback, it is important to use a lens with as large a numerical aperture as possible (to make β large) and to apply low-reflectivity AR coatings. The single-pass output slope efficiency, SE, is given by $SE = FSE \cdot T_L \cdot R_{g0}$, where FSE is the diode front-facet slope efficiency and R_{g0} is the grating reflectivity in zeroth order. Since the grating only diffracts in first order, any increase in the output coupling is accompanied by a decrease in feedback (assuming that grating scattering losses are low). This is illustrated in Fig. 1, where light-power vs current (L–I) curves are shown for three gratings with different diffraction efficiencies. For the grating giving the highest output efficiency, curve (A), threshold is not much reduced below that of the diode without feedback. This grating is specially chosen to give low feedback due to the small blaze angle of 14° (blaze wavelength = 4000 Å). From the measured values of R_{g1} and threshold changes, we estimate a coupling efficiency of 0.5.

A general property of gratings is that the diffraction efficiency for S-polarized light (electric

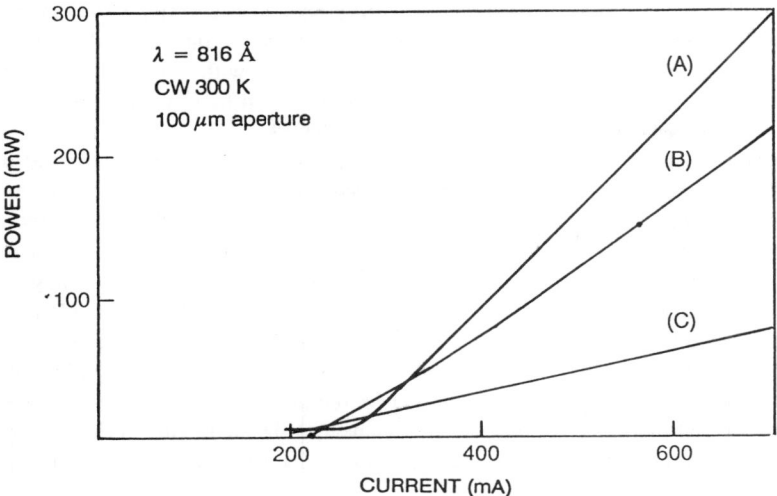

Fig. 1. Continuous output power vs drive current for the external-cavity laser with feedback from three different gratings. 1200 g/mm ruled grating, 4000 Å blaze, Au coated, SE = 0.67 mW/mA (A); 1200 g/mm holographic grating, Al coated, SE = 0.46 mW/mA (B); 1880 g/mm holographic grating, Au coated, SE = 0.15 mW/mA (C).

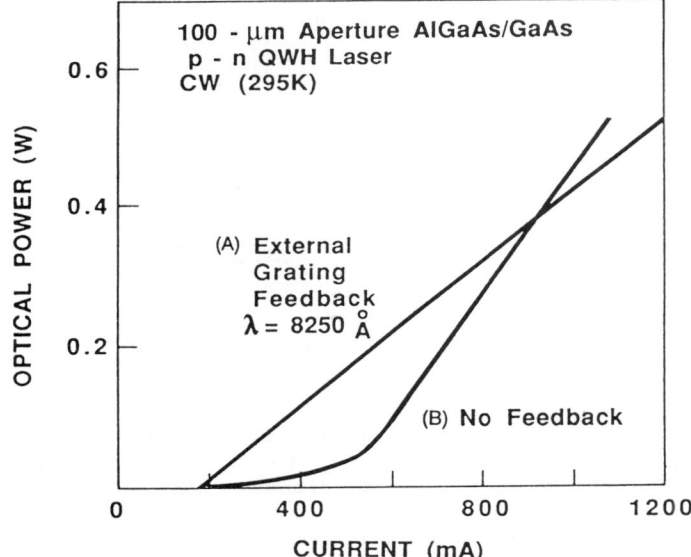

Fig. 2. Plot of the CW power coupled out from the external-cavity laser vs drive current at a wavelength of 8350 Å (curve A), and a plot of the diode power output with no external feedback (curve B). The output slope efficiency of the external-cavity laser is 0.51 W/A.

field vector **E** parallel to the grooves) is higher than that for the orthogonal P polarization. This has been confirmed both experimentally and theoretically.[10] The polarization of the external-resonator laser diode emission is selected by competition between diode facet reflectivity and grating reflectivity. As is well known, the facet reflectivity is higher for TE modes than for TM modes. Experimentally, we observe that the laser emission is always polarized with E parallel to the grooves, which corresponds to TE polarization in the cavity geometry used here, which indicates that the residual facet reflectivity is selecting the polarization.

A CW light output vs current (L–I) plot of the external resonator laser output at a wavelength of 8250 Å is shown in Fig. 2 (curve A), together with an L–I plot of the diode without feedback (curve B). The threshold current is reduced by more than a factor of 2 by grating feedback, showing the high grating-diode coupling that is obtained. The L–I curve of the diode output with external grating feedback is linear, with a slope of 0.51 W/A. This slope efficiency is consistent with the measured values of $R_{g0} = 0.57$, $T_L = 0.95$, and FSE = $1.05 \div 1.10$ W/A. A power of 550 mW is coupled out of the cavity at 1.2 A. This is a factor of 7 greater than the CW power output reported by Epler et al.[7] The power output is limited by catastrophic optical damage to the laser facet, which occurs at ~ 1.5 A of drive current for the diodes used in the present experiment. The linearity of the L–I plot indicates that no misalignment of the cavity due to thermal drift takes place as the current is ramped, and that power fluctuations due to mode jumps are small.

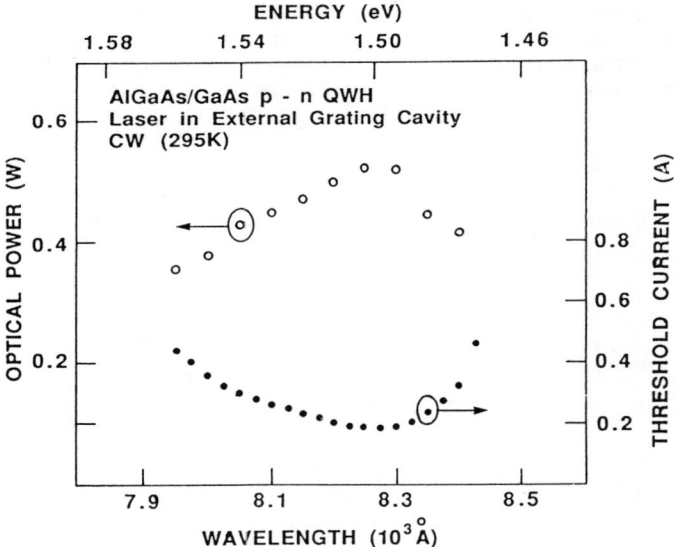

Fig. 3. CW threshold current (full circles) and output power (open circles) of the external-cavity laser as a function of grating-tuned wavelength. The output power is measured at a drive current of 1200 mA.

The dependence of threshold current and output power at a fixed drive current of 1.2 A on the grating-tuned lasing wavelength are shown in Fig. 3. The tuning range is about 500 Å, from 7950 to 8450 Å. At either end of the tuning curve, the feedback from the grating is insufficient to lock the diode emission onto a single wavelength, and multiple Fabry–Perot modes appear in the spectrum, i.e., the diode lases due to its residual facet reflectivity. Tuning ceases when the threshold of the diode with feedback exceeds the bare-laser threshold. Therefore, in order to increase the tuning range, it is necessary to raise the threshold by applying as low a reflectivity coating as possible.

The variation of power with wavelength is relatively small, with a maximum decrease of ~ 25% from the peak at the extremes of the selected tuning range. The minimum threshold current of ~ 190 mA occurs in a wavelength range between 8200 and 8250 Å, which is 200 Å longer than the lasing wavelength of the diode without feedback. The two curves in Fig. 2 are constructed by aligning the cavity initially at the short-wavelength end of the tuning range, tuning, and subsequently repeating the alignment at the center of the tuning range. A similar set of curves can be traced by adjusting the cavity at the center of the tuning range only, and then tuning, at the expense of about a 10% reduction in output power. Due to the residual reflectivity of the AR-coated output facet, the power varies slightly as the wavelength is tuned through Fabry–Perot modes of the diode. However, the peak-to-peak variation is under 5%, and tuning is continuous between Fabry–Perot modes. The output slope efficiency is a monotonically increasing function of the lasing wavelength. A maximum efficiency of 0.55 W/A is measured at 8400 Å, on the long-wavelength limit of the tuning range. This increase in efficiency with tuned wavelength is due to the reduction in internal absorption losses as the laser line is tuned to the low-energy side of the gain peak.

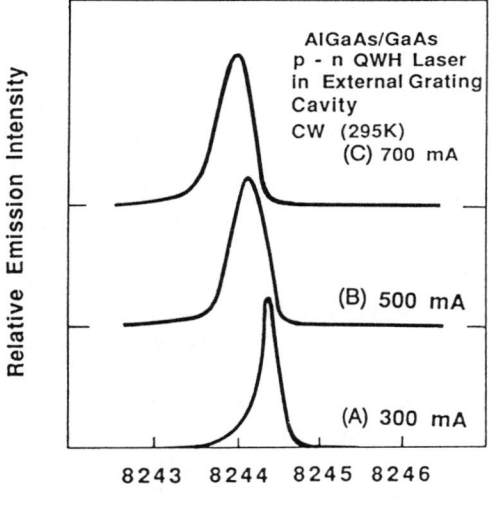

Fig. 4. Spectral width of the external-cavity laser output as a function of diode current. The linewidth $\Delta\lambda$ at half-maximum (corrected for spectrometer resolution) is < 0.2 Å in (A), 0.4 Å in (B), and 0.5 Å in (C). The power corresponding to the current indicated on curves (A)–(C) is 55, 160, and 260 mW, respectively.

The emission spectrum at three different drive currents, measured with a 0.5-m grating spectrometer, is shown in Fig. 4 (curves A–C). No adjustments have been made to the cavity between measurements. At 300 mA (1.5 I_{th} at this wavelength, curve A the apparent linewidth at half-maximum intensity is 0.2 Å, which is equal to the measured resolution of the spectrometer. The structure in the spectrum displayed in curve A is an artifact caused by diffraction at the spectrometer slits. The linewidth increases at higher drive currents to 0.4 Å at 500 mA and 0.5 Å at

700 mA (curves B and C, respectively). At the maximum output power, the linewidth increases to 1 Å. In addition to exhibiting spectral broadening, the output emission shifts to shorter wavelength by 0.4 Å as the diode current is increased from 300 to 700 mA. This small spectral shift is probably due to thermal expansion of the diode heat sink. At a fixed current, the wavelength is stable to within 0.05 Å for a period of tens of minutes. The spectral width can be estimated from an expression that takes into account the linear dispersion and resolving power of the grating-lens system:[2,9]

$$\Delta\lambda = \frac{d\lambda}{dx} \cdot \Delta x_m, \qquad (1)$$

where $\Delta\lambda$ is the wavelength spread of the light coupled back into the active region, $d\lambda/dx$ is the linear dispersion on the focal plane of the lens, and Δx_m is the mode width perpendicular to the p-n junction. The dispersion of a Littrow-mounted grating coupled with a collimating lens of focal length f is given by

$$\frac{d\lambda}{dx} = \frac{\lambda}{2f \cdot \tan\theta}, \qquad (2)$$

where θ is the angle between the grating normal and the incident beam. With $\theta = 30°$, $f = 6.5$ mm, $\lambda = 0.82$ μm, and $\Delta x_m \approx 1$ μm, the value of $\Delta\lambda$ obtained from equations (1) and (2) is ~ 1 Å. The measured linewidth is smaller than the estimated one at lower output powers, probably due to inhomogeneous gain saturation. However, at the maximum output power, the linewidth is consistent with the grating and lens resolution, which is in agreement with the results obtained by Rossi et al. a at pulsed power of 3.5 W.[2] The spectra shown in Fig. 4 display a 5-Å interval cente ' on the lasing line, and would thus include only 2 to 3 Fabry–Perot modes of the laser diode chip. In spectral scans over a wider 50-Å range around the lasing line, no Fabry–Perot modulation is detected at a sensitivity of 10^3 times that used for the spectra in Fig. 4. This result indicates that the overall gain is saturated by grating feedback.[11]

In order to get more highly resolved spectra of the laser emission, we have measured the output with a scanning spherical-mirror interferometer. In contrast to a confocal interferometer, the free spectral range of this device is variable. In particular, it is practical to operate with mirror separations of several hundred microns, giving a free spectral range as large as 1000 GHz, which is useful for relatively broad linewidth sources. Typical spectra are shown in Fig. 5., illustrating the increase in linewidth as the diode is driven further above threshold. At all currents, the output spectrum consists of numerous axial modes of the external resonator, which are spaced 1.2 GHz apart. Just above threshold, as shown in the uppermost curve of Fig. 5, the diode emission consists essentially of a single external-cavity mode that rises above a background of 4–5 lower-intensity modes that span ~ 6 GHz. As the current is increased, the background modes grow in intensity and the line center shifts to higher frequency (to the right in Fig. 5).

The spatial profile and divergence of the output beam are important in applications such as pumping of Nd-doped solid-state lasers, which is described in the second part of the paper. In Fig. 6 we show measured beam profiles perpendicular and parallel to the active layers, 150 mm from the grating, at an output power of 160 mW ($2.6 I_{th}$, $\lambda = 8250$ Å). The perpendicular profile (Fig. 6(a)) can be fitted to a truncated Lorentzian shape[12] with a width equal to the 8-mm aperture of the collimating lens. In this direction, the beam converges to a waist at a distance > 10 m from the grating. The profile does not change shape with diode current, which is to be

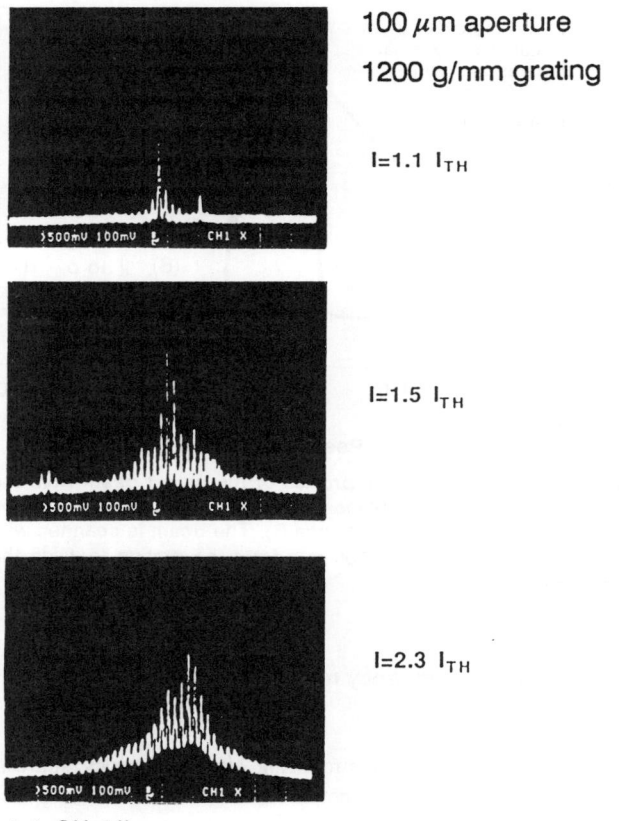

Fig. 5. Spectrum of external-cavity laser as measured with a scanning interferometer having a free spectral range of 190 GHz and a finesse > 1000.

expected from the real refractive index guiding in the diode perpendicular to the *p-n* junction. In the parallel direction (Fig. 6 (A)), the beam profile is sensitive to the tilt of the grating. However, when the grating tilt is adjusted to maximize the output power, a profile similar to that shown in Fig. 6 (B) is always observed. The beam divergence corresponding to the spatial profile in Fig. 6 (B) is 0.37° (full angle at half-maximum intensity). At higher currents, the parallel divergence is approximately 0.4°. This value of divergence is comparable to what has been obtained with wide-stripe gain-guided lasers in nonselective external cavities.[13,14] It is consistent with the divergence of a 100-μm wide incoherent source at the focal point of a 6.5 mm-focal length lens.

In this section, we have described a high-power, tunable, narrow-linewidth external cavity diode laser. This device has a low-divergence output beam and a high efficiency, making it useful for applications such as pumping of rare-earth doped solid-state lasers and second harmonic generation.

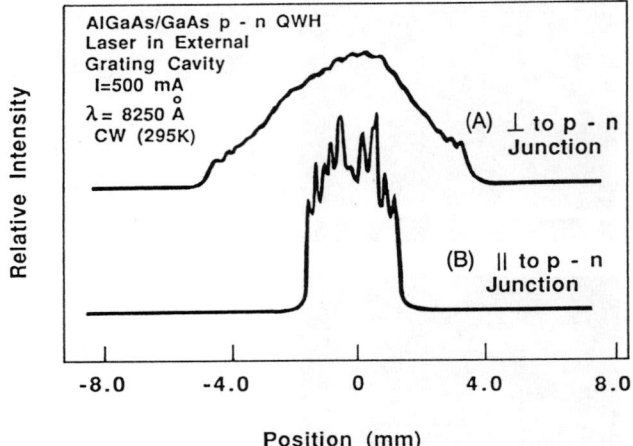

Fig. 6. Transverse spatial profiles of the output beam from the external cavity laser in planes perpendicular to the *p-n* junction (curve A) and parallel to it (curve B). The beam is scanned with a 100-μm slit at a point 150 mm from the grating (outside the resonator).

3. Temperature-tuned single-frequency microcavity lasers fabricated from flux-grown Nd,Ce:YAG

Diode-pumped Nd microcavity lasers are promising single-frequency sources at 1.06 μm and 1.34 μm because of their simplicity and compactness. In contrast to more complicated single-frequency nonplanar-ring laser designs, microcavity lasers do not require precise matching of pump beam size to the lasing mode and are simple to align. An initial report has demonstrated a linewidth as narrow as 50 kHz and single-frequency CW operation at a pump power over 20 times threshold.[15] In these lasers, the requirement for single axial mode operation is that the cavity length L be sufficiently short for the axial mode spacing $\Delta\lambda$ to be larger than the gain bandwidth of the medium. The mode spacing is calculated from the standard expression $\Delta\lambda = \lambda^2/2nL$, where n is the refractive index of the crystal. For the 1.06-μm line of Nd^{3+} in yttrium aluminum garnet (YAG), the gain bandwidth at 300 K is 6 Å,[16] which yields a maximum cavity length of 520 μm. A cavity of this short results in incomplete absorption of diode pump light for standard Czochralski-grown YAG crystals doped with 1% Nd, which have a peak absorption of 8 cm^{-1} at a wavelength of 8086 Å. It is impractical to increase the Nd concentration much beyond 1% because of segregation in the melt, leading to large nonuniformities in Nd doping. However, by using the flux-growth technique, it is possible to grow uniformly doped crystals containing as much as 4% Nd.[17]

In the present work we report on a diode-pumped single-frequency microchip laser fabricated from flux-grown YAG crystals containing 2.4% Nd. The laser emits 45 mW in a single axial mode in CW operation. The output wavelength is tunable over a range of 6 Å, at a center wavelength of 1.064 μm, by changing the laser temperature.

The laser is fabricated by polishing plane-parallel resonator facets on the YAG crystal and depositing dielectric mirrors directly on these facets. The cavity length is 540 μm. The measured input mirror reflectivities are 0.998 at 1.06 μm and 0.05 at 0.809 μm; the corresponding values for the output mirror are 0.95 and 0.99, respectively. The microchip is pumped with the focused output of the high-power grating-tuned diode laser described in the previous section. The focusing optics consists of a 120 mm-focal length spherical lens followed by a 5 mm-focal length cylindrical lens. The pump spot incident on the Nd laser is slightly elliptical, measuring 50 μm by 70 μm.

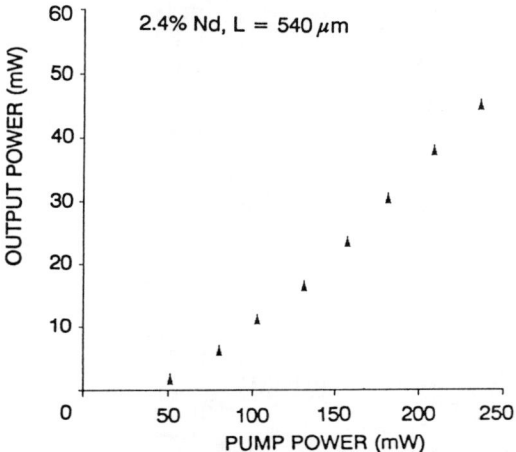

Fig. 7. Continuous output power of single-frequency Nd:YAG microlaser as a function of incident pump power. The pump source is the external-cavity laser tuned to a wavelength of 8086 Å. The pump beam is focused into a spot measuring 50 μm × 70 μm. The slope efficiency is 25% after correcting for pump transmission.

The CW output power as a function of pump power is shown in Fig. 7. Threshold is reached at 50 mW, and the output power is linear with pump power above $2P_{th}$. The laser lases in a single axial mode up to the maximum output power of 45 mW. The slope efficiency is 0.25. From the slope efficiency, we estimate the absorption losses in the crystal (α_i) to be 1.2 cm^{-1}, using the following expression for end-pumped laser crystals:

$$\eta_S = \frac{\nu_0}{\nu_p} \frac{\ln(1/R_0)}{\ln(1/R_0) - 2L\alpha_i}$$

Here ν_0 and ν_p are the lasing and pump frequencies, respectively, and R_0 is the reflectivity of the output mirror. This expression is valid provided that the pump beam is smaller than the lasing mode, which is the case in our experiments. The high loss in the present crystal is probably due to impurities incorporated from the flux.

The output beam profile is shown in Fig. 8. The device lases in a TEM$_{00}$ mode with a slightly elliptical spot. In the direction parallel to the long axis of the pump spot, the waist diameter is 105 μm, and in the perpendicular direction the waist diameter is 97 μm. It has been established previously that the thermally-induced refractive index gradient from pump light absorption is responsible for mode guidance in microchip lasers.[18] In the present experiment, thermal diffusion has broadened the profile of the pump beam sufficiently so that the lasing mode is only slightly asymmetrical. As expected from thermal guiding, the spot size decreases slightly with increasing pump power, but the profile remains Gaussian.

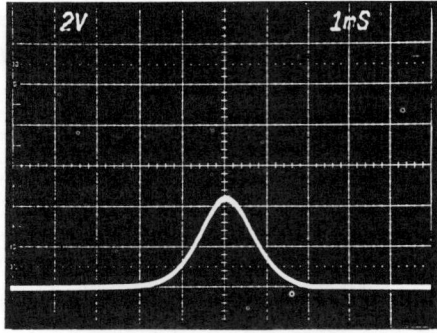

$\theta_\perp$ = 13.9 mrad

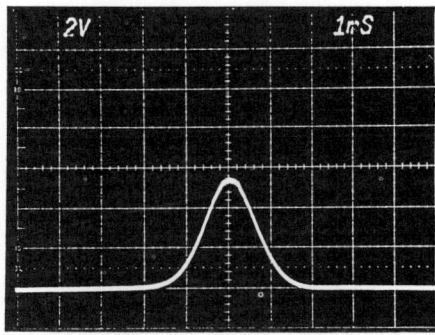

$\theta_\parallel$ = 13.9 mrad

5.5 mrad/div

Fig. 8. Spatial profile of the Nd:YAG microlaser output beam at 1.4 P_{th}. The beam is Gaussian with a slightly elliptical waist measuring 97 μm ×105 μm. L = 540 μm.

The position of the cavity axial modes relative to the gain peak can be controlled by varying the microlaser temperature. As the axial mode traverses the gain, the threshold will change. This behavior is clearly seen in Fig. 9, which slows the threshold pump power as a function of crystal temperature. At room temperature, the wavelength of the axial mode coincides with the peak of the gain, and the threshold is at a minimum. This is confirmed by the lasing spectrum which shows a single axial mode located at the peak of the fluorescence intensity. Throughout the entire temperature range, the laser operates in a single frequency.

Temperature tuning of the single-frequency output is shown in Fig. 10. The lasing wavelength can be tuned over more than 6 Å in a 60-degree temperature interval. This tuning range is limited by the axial mode spacing—i.e., a neighboring mode enters the gain curve as the temperature is changed to increase tuning. With a shorter cavity, the tuning range can be further increased. The temperature coefficient of tuning is 0.12 Å/°C. The change in lasing wavelength depends on both

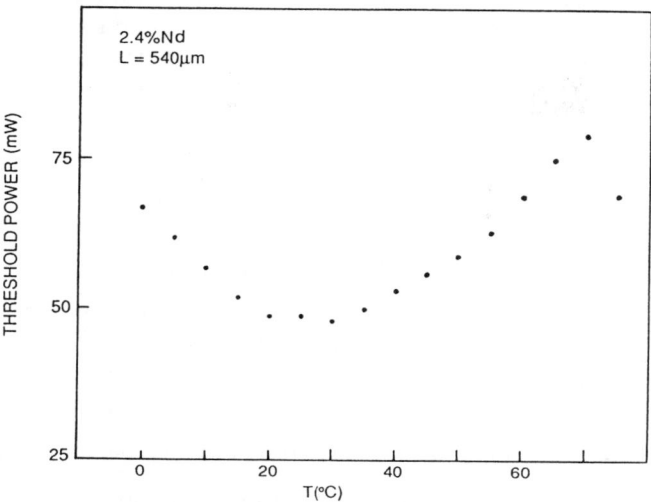

Fig. 9. Variation of threshold pump power with Nd:YAG crystal temperature.

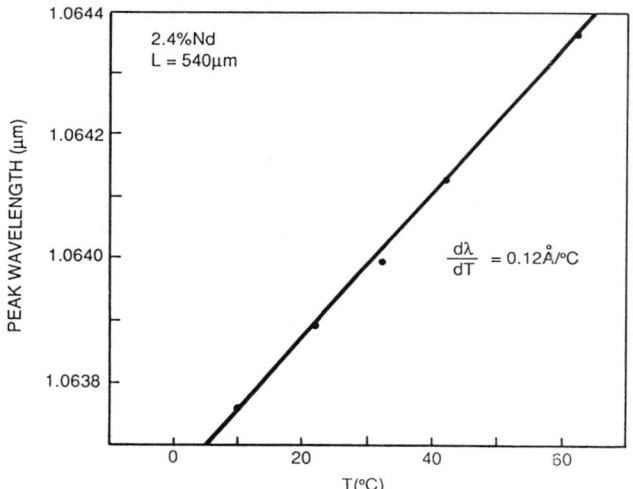

Fig. 10. Lasing wavelength as a function of temperature at an output power of 10 mW.

the resonator thermal expansion coefficient and on the temperature variation of the refractive index, as expressed by the following relation:

$$\frac{d\lambda}{dT} = \lambda \left(\frac{1}{L} \cdot \frac{dL}{dT} + \frac{1}{n} \cdot \frac{dn}{dT} \right)$$

Using published values for the thermal expansion coefficient and temperature dependence of the refractive index of YAG, we obtain $d\lambda/dT = 0.11$ Å/°C, in close agreement with the measured value.

In summary, we have described a temperature-tuned single-frequency Nd:YAG monolithic-cavity laser. The maximum output power of 45 mW is limited by the available pump power, and not by the performance of the laser. With higher pump power and higher output coupling, it should be possible to obtain several hundred milliwatts of single-frequency output.

References

[a] Visiting Scientist on leave from A.F. Ioffe Physico-Technical Institute, Academy of Sciences of the USSR, Leningrad, USSR
[1] J.A. Rossi, S.R. Chinn, and H. Heckscher 1973 *Appl. Phys. Lett.* **23** 25
[2] H. Heckscher and J.A. Rossi 1975 *Appl. Opt.* **14** 94
[3] J.A. Rossi, J.J. Hsieh, and H. Heckscher 1975 *IEEE J. Quantum Electron.* **QE-11** 538
[4] V.Yu. Bazhenov, A.P. Bogatov, P.G. Eliseev, O.G. Okhotnikov, G.T. Pak, M.P. Rakhvalsky, M.S. Soskin, V.B. Taranenko, and K.A. Khairetdinov 1985 *IEE Proc. J.* **132** 9
[5] R. Wyatt and W.J. Devlin 1983 *Electron. Lett.* **19** 110
[6] J.E. Epler, N. Holonyak, Jr., R.D. Burnham, C. Lindstrom, W. Streifer, and T.L. Paoli 1983 *Appl. Phys. Lett.* **43** 740
[7] J.E. Epler, G.S. Jackson, N. Holonyak, Jr, R.L. Thornoton, R.D. Burnham, and T.L. Paoli 1985 *Appl. Phys. Lett.* **47** 779
[8] J.E. Epler, N. Holonyak, Jr., J.M. Brown, R.D. Burnham, W. Streifer, and T.L. Paoli 1984 *J. Appl. Phys.* **56** 670
[9] J.E. Lawler, W.A. Fitzsimmons, and L.W. Anderson 1976 *Appl. Opt.* **15** 1083
[10] E.G. Loewen, M. Neviere, and D. Maystre 1977 *Appl. Opt.* **16** 2711
[11] H. Bachert, A.P. Bogatov, P.G. Eliseev, A. Keiper, and K.-A. Khairetdinov 1979 *IEEE J. Quantum Electron.* **QE-15** 786
[12] A. Naqwi and F. Durst 1990 *Appl. Opt.* **29** 1780
[13] C.J. Chang-Hasnain, J. Berger, D.R. Scifres, W. Streifer, J.R. Whinnery, and A. Dienes 1987 *Appl. Phys. Lett.* **50** 1465
[14] W.F. Sharfin, J. Seppala, A. Mooradian, B.A. Soltz, R.G. Waters, B.J. Vollmer, and K.J. Bystrom 1989 *Appl. Phys. Lett.* **54** 1731
[15] J.J. Zayhowski and A. Mooradian 1989 *Opt. Lett.* **14** 24
[16] T. Kushida 1969 *Phys. Rev.* **185** 500
[17] A.A. Kaminskii 1990 *Laser Crystals, 2nd ed. (Springer–Verlag Berlin, 1990)* p. 105.
[18] J.J. Zayhowski 1990 *Proccedings of the Optical Society of America Topical Meeting on Advanced Solid State Lasers, Salt Lake City, Utah* March 1990

The influence of leakage on the characteristics of QW Lasers

V.B. Khalfin, A.B. Gulakov, I.V. Kochnev, E.U. Rafailov, Yu.M. Shernyakov, B.S. Yavich, and D.Z. Garbuzov

A.F. Ioffe Physico-Technical Institute, Academy of Sciences of the USSR,
26 Polytekhnicheskaya st. 194021 Leningrad, USSR

Abstract. The paper reports on theoretical and experimental studies of the threshold current density and differential quantum efficiency dependencies on output losses for AlGaAs/GaAs and InGaAsP/GaAs SQW SCH laser diodes. A theoretical model is proposed to calculate the effect of waveguide recombination and leakage to the cladding on the threshold current and differential quantum efficiency of SQW SCH lasers. It is shown that the model assuming quasineutrality and continuity of quasi Fermi levels at interfaces gives a correct description of the process of carrier recombination in the waveguide layers of the lasers in question. Carrier leakage from the active region should be the main cause of increase of the threshold current density in the SQW lasers for densities in excess of ~ 3 kA/cm^2. It has been established that the carrier concentration in the waveguide increases above the lasing threshold as well, which can result in enhanced leakage into the claddings with increasing current density and to an anomalous decrease of differential efficiency in short-cavity laser diodes for structures with small ($\lesssim 100$ meV) bandgap difference between the waveguide and claddings layers. The experimentally revealed decrease of the differential efficiency in short-cavity diodes is essentially faster than that predicted by the model. One of possible causes of the decrease of differential efficiency is the enhanced filamentation of lasing with decreasing cavity length.

The present paper discusses the results of the theoretical and experimental studies into the dependences of the threshold current density (J_{th}) and differential quantum efficiency (η_d) on output losses (α_{out}) in SQW SCH laser diodes based on InGaAsP/GaAs and AlGaAs/GaAs heterostructures which were grown by modified LPE[1] and MOCVD,[2] respectively. In contrast to DH lasers, in QW laser diodes the growth of the threshold current density with increasing losses is known to become superlinear. Basically, this phenomenon is the result of gain saturation in the quantum well. However some studies[3-5] dealing with AlGaAs/GaAs SQW SCH lasers revealed an increase of threshold current density with losses which is stronger than it would follow from the theory taking into account only the gain saturation effect, and showed that waveguide recombination and carrier leakage through heterobarriers to the cladding layers can affect substantially the behavior of the $J_{th} = f(\alpha_{out})$ dependence in these lasers. Similar phenomena were observed to occur also in InGaAsP/InP lasers[6] emitting at $\lambda = 1.3$ μm. As shown in our preceding work,[7] current dependent leakages in InGaAsP/InP ($\lambda=1.3$ μm) SQW SCH lasers result not only in a superlinear growth of the threshold current density but in a decrease of the differential quantum efficiency with increasing output losses as well.

For AlGaAs/GaAs SQW SCH lasers emitting at 0.85 μm the drop in differential efficiency with decreasing cavity length revealed by Wilcox et al.[3] was attributed to absorption by non-equilibrium free carriers in the waveguide. This explanation is hardly plausible, since the density of non-equilibrium carriers in the waveguide required to account for the drop of differential efficiency in the very short cavity lasers should be about 10^{19} cm^{-3}. Such a high concentration of non-equilibrium carriers should result in the onset of lasing in the waveguide which was not observed in the experiment. Apart from this, the corresponding estimates of the current density yield values which exceed by far those observed experimentally even if only radiative recombination in the waveguide is included in the calculations.

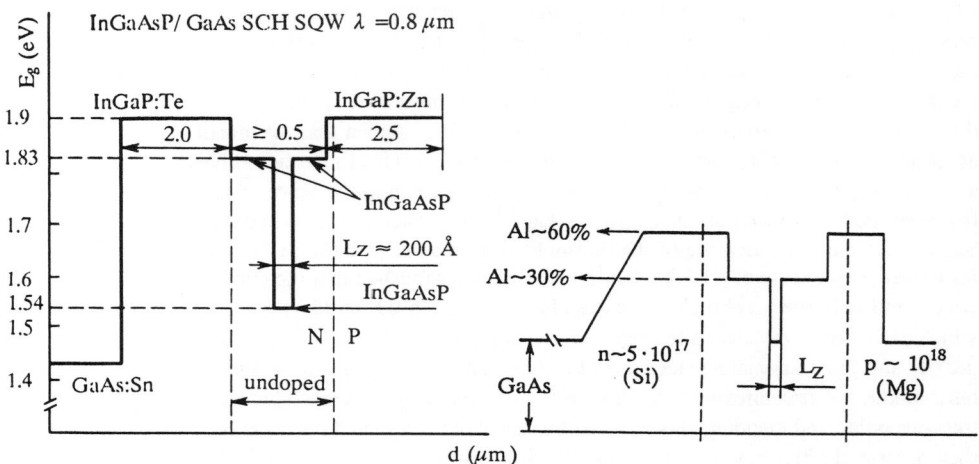

Fig. 1. Schematic diagram of (a) InGaAsP/GaAs and and (b) AlGaAs/GaAs laser structures. L_z is active layer thickness.

This study was performed on laser diodes fabricated of the heterostructures whose band diagrams are presented in Fig. 1(a,b). Note that InGaAsP/GaAs structures differ from the AlGaAs/GaAs structures in a very small bandgap difference ($\Delta E_g \approx 70$ meV) between the waveguide and cladding layers. The InGaAsP/GaAs structures had an active region ~ 200 Å thick.

The active region thickness in the AlGaAs/GaAs structures was 60, 100, and 200 Å. These structures were used to prepare broad area lasers with stripes 20–100 μm wide, the insulation being provided by an oxide film or a Schottky barrier. The dependences of the threshold current and differential quantum efficiency near the lasing threshold on cavity length were measured. It was found that the threshold current density for a given cavity length decreases slightly with increasing stripe width within 20 to 60 μm. Since, however, the results obtained on diodes with the stripe width 60 and 100 μm were fairly close, the main measurements were performed on the 100 μm-stripe diodes. The cavity length in the diodes studied varied from 50 μm to 2 mm. The output losses were also varied by depositing antireflection and reflective coatings on the diode facets.

In the case of InGaAsP/GaAs-based diodes with the same cavity length taken from different wafers one observed a fairly large scatter between the values of both the threshold current

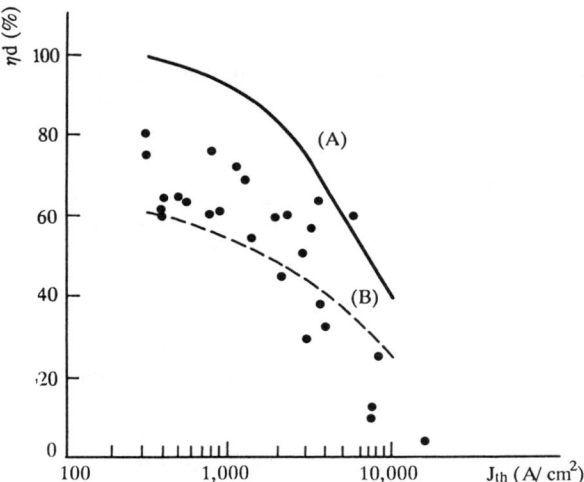

Fig. 2. Differential quantum efficiency (η_d) vs threshold current density J_{th} for InGaAsP/GaAs laser diodes. Points are experimental values, solid line (A) represents calculated relation $\eta_e = f(J_{th})$, dashed line (B), evaluation of the lower limit η_d' ($\eta_d' = 0.6\eta_e$).

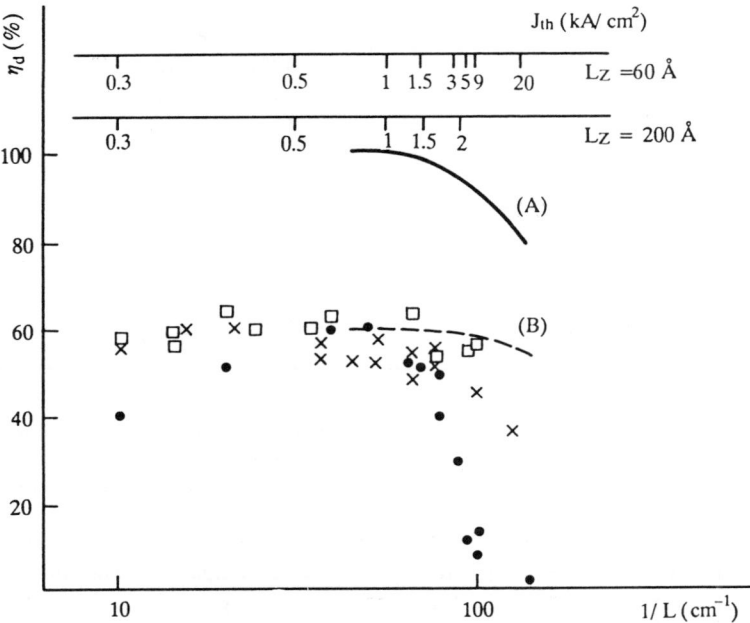

Fig. 3. Differential quantum efficiency vs reciprocal cavity length for AlGaAs/GaAs laser diodes. The two upper scales refer to the threshold current densities for diode active region thicknesses of 60 Å and 200 Å. The squares, stars and circles identify experimental values for diodes with L_z = 200, 100 and 60 Å. Calculated curves A and B relate to diodes with L_z = 60 Å. Solid curve displays $\eta_e(\alpha_{out})$, and dashed curve, $\eta_d(\alpha_{out})$.

density and differential quantum efficiency. Therefore Fig. 2 presents for these diodes the dependence of the differential quantum efficiency not on output losses but rather on the threshold current density. For AlGaAs/GaAs diodes with active regions 60, 100, and 200 Å thick the conventional $\eta_d = f(\alpha_{out})$ relations are given in Fig. 3. The additional scales in the upper part of the figure specify the threshold current density for diodes with active region thicknesses of 60 Å and 200 Å.

To reveal the mechanisms responsible for the laser characteristics, additional measurements of the quantum efficiency of spontaneous emission from the active region (η_a) and the waveguide (η_w) were made as a function of current density. The experiments were carried out on specially prepared samples with a ~ 5 μm-wide slit in the upper stripe contact through which the radiation was coupled out. The integral intensity for the emission bands associated with spontaneous radiative recombination in the active region and the waveguide was measured. The measurements were performed both on lasing diodes, below and above the lasing threshold, and on very short-cavity samples, 50–70 μm long, in which the lasing did not set in at the current densities investigated. On very short cavity samples one measured the intensity of spontaneous emission when coupled out through the diode facet as well. Measurements with output coupling through the facet permit also evaluating the absolute quantum efficiency of spontaneous emission. According to these measurements, the highest values of η_a reached at current densities $(0.5-1) \times 10^3$ A/cm^2 correspond to a close to 100% internal efficiency of radiative recombination for both diode types studied. The behavior of the $\eta_a = f(J)$ relations did not depend on the actual direction of observation.

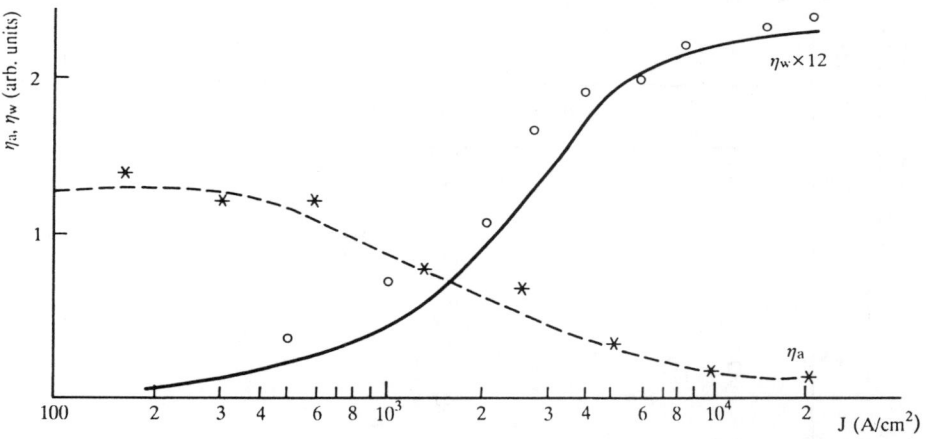

Fig. 4. Quantum efficiency of spontaneous emission from active region (η_a) and waveguide (η_w) vs current density J for very short-cavity (L $\simeq$ 50 μm) non-lasing InGaAsP/GaAs diodes. Stars and circles are experimental values of η_a and η_w. Solid and dashed curves represents calculated relations.

Figs. 4 and 5 present the measured relative values of η_a and η_w for very short-cavity, non-lasing InGaAsP/GaAs and AlGaAs/GaAs diodes with output coupling through the slit. The decrease of the quantum efficiency for the active region at low current densities can be accounted for by non-radiative recombination via deep levels, and the drop in η_a at high densities, by recombination of non-equilibrium carriers in the waveguide and leakage to the cladding, which is supported by the growth of η_w in the corresponding region of current densities.

Shown in Figs. 6 and 7 are the results obtained on the lasing samples. One can see that in the case of long cavity diodes characterized by comparatively low threshold current densities (long cavity diodes) the intensity of spontaneous emission from the active region saturates after the onset of lasing, whereas that from the waveguide continues to grow, although at a slower rate than it did before the lasing. For short cavity samples with a high lasing threshold ≥ 2 kA/cm^2 the intensity of spontaneous emission from the active region continues to increase after the onset of lasing as well, thus indicating the absence of Fermi level pinning above the lasing threshold, at least in a part of the active region area. This behavior is apparently associated with filamentation of lasing which is confirmed by the strongly non-uniform near-field pattern typical of short-cavity lasers.

We attempted to interpred the results of the experiments presented in Figs. 2–7 in terms of a model including waveguide recombination of non-equilibrium carriers in the waveguide and leakage to the cladding. The calculations by this model were performed under the following simplifying assumptions:

(*i*) All layers in the structure (active region, waveguide, cladding) are quasineutral, the thickness of the space charge layers in the waveguide constitues only a small fraction of its total thickness, so that recombination in them may be neglected.

(*ii*) The electron and hole quasi Fermi levels are continuous at the interfaces.

(*iii*) Spatial quantization is included only in calculations of the quasi Fermi level positions and radiative recombination rate in the active region.

(*iv*) The radiative recombination rate in the active region is calculated in terms of the simple parabolic band model with selection rules for the quasimomentum and subband number, the contribution of the non-radiative processes being neglected.

(*v*) Wavequide recombination for InGaAsP/GaAs and AlGaAs/GaAs diodes was calculated by different models. Indeed, for LPE-grown InGaAsP/GaAs structures there is an experimental evidence[1] for the high efficiency of radiative recombination in all layers of the structure. On the contrary, measurements carried out on MOCVD-grown AlGaAs/GaAs structures reveal a high rate of non-radiative recombination in their waveguide layers.[8] Therefore for the waveguide of the InGaAsP/GaAs diodes we took into consideration only the radiative recombination, and for that in the AlGaAs/GaAs diodes, linear-in-concentration non-radiative, and quadratic radiative recombination.

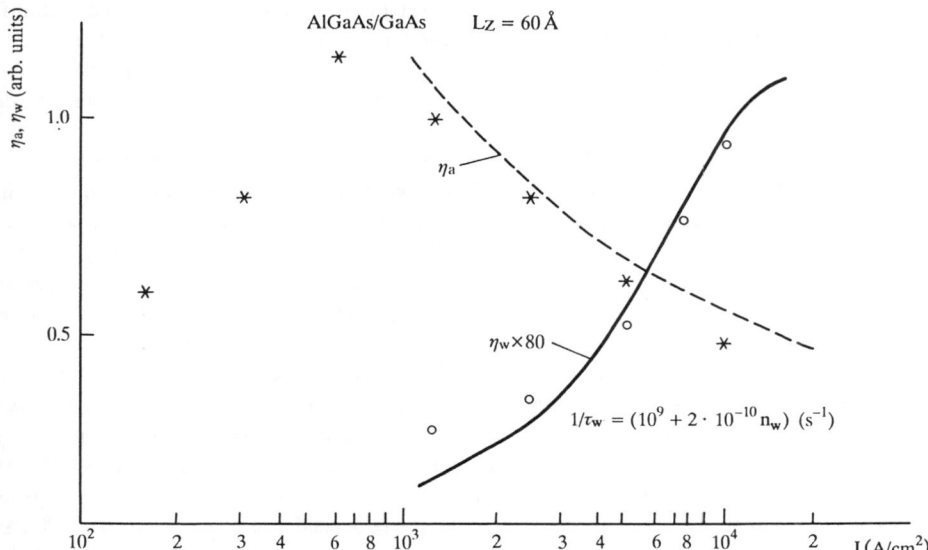

Fig. 5. Same as in Fig. 4, but for AlGaAs/GaAs diodes with L_z = 60 Å.

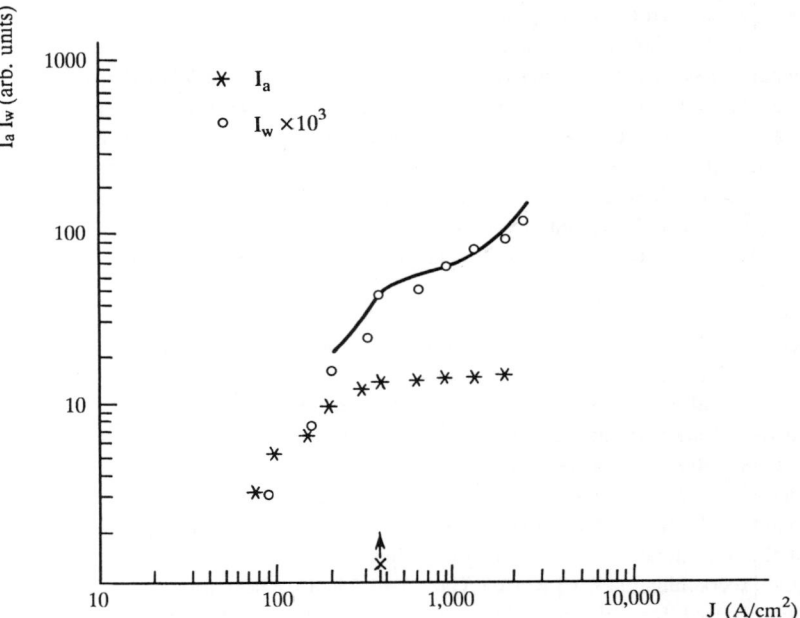

Fig. 6. Spontaneous emission intensity from active region (I_a) and waveguide (I_w) vs current density J for an InGaAsP/GaAs laser diode with cavity length L = 1.3 mm. Stars and circles: experimental values of I_a and I_w. Solid curve: calculated $I_w = f(J)$ relation. The curve is normalized to experimental values which were obtained at the threshold current density identified by the arrow.

(*vi*) In the waveguide there is a gradient of non-equilibrium hole concentration associated with diffusive hole transport from the cladding to the active region. The magnitude of this gradient is calculated from the ambipolar diffusion equations with the inclusion of recombination in the waveguide. By virtue of quasineutrality, the electron concentration should also increase in the waveguide from the active region to the p-cladding.

(*vii*) Leakage to the n-cladding is neglected, and one considers only electron leakage to the p-cladding calculated by the conventional expression[7] taking into account the diffusion of non-equilibrium carriers and their drift in the Ohmic field across the p-cladding of a finite thickness.

(*viii*) In the calculation of the effect of leakage on differential quantum efficiency it was assumed that above the lasing threshold the quasi Fermi levels are pinned in the active region, the average non-equilibrium carrier concentration in the waveguide growing approximately proportional to the recombination current in the quantum well.

Let us turn now to a comparison of the calculations with experimental data. The calculations of η_a and η_w for the InGaAsP/GaAs and AlGaAs/GaAs diodes vs current density are shown in Figs. 4 and 5 with solid lines. We readily see that in the region of maximum current density the quantum efficiency of emission from the waveguide should saturate for the InGaAsP/GaAs diodes, whereas for the AlGaAs/GaAs diodes in this region of current η_w should continue to grow with current. The saturation of the calculated current density dependence of η_w for InGaAsP/GaAs laser diodes is connected with electron leakage to the p-cladding. The good agreement of the calculations with the experimental data suggests that the difference between the $\eta_w = f(J)$ relations for the InGaAsP/GaAs and AlGaAs/GaAs diodes is caused by the small height of the cladding/waveguide barrier in InGaAsP/GaAs structures.

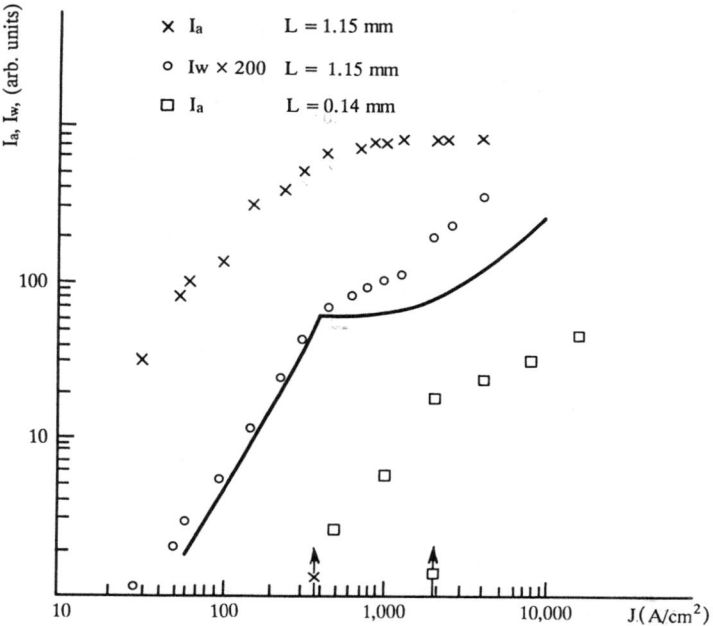

Fig. 7. Same as Fig. 6 for AlGaAs/GaAs diodes with a cavity length $L \approx 1.15$ mm (circles and stars) and $L \approx 140$ μm (squares) prepared from a structure with active region 130 Å thick and Al content in the waveguide and cladding layers of 20 and 38 at %. Threshold current densities for both diodes are identified by arrows with corresponding symbols. Solid curve: calculation of I_w for diode with $L \approx 1.15$ mm.

The calculations characterizing the dependence of the differential quantum efficiency on threshold current density for InGaAsP/GaAs structures are shown in Fig. 2 with a solid and a dashed lines. The solid curve was calculated as the ratio of the increment of the recombination current in the quantum well due to stimulated emission (ΔJ_{QW}) to that of the total current (ΔJ). The values of $\eta_\ell = \Delta I_{QW}/\Delta J$ correspond to the upper limit of differential quantum efficiency determined only by the current leakage. The differential quantum efficiency (η_d) is equal to the product of η_ℓ by the ratio of the output to total losses, $\eta_{out} = \alpha_{out}/(\alpha_{in} + \alpha_{out})$. As seen from Fig. 2, for long-cavity lasers with low threshold current densities (≤ 1 kA/cm^2) which correspond to diode cavity lengths of 2 mm to 400 μm, the values of differential quantum efficiency lie typically between 0.6 and 0.8. Since within this range of currents $\eta_\ell \approx 1$, the differential quantum efficiency in this range is totally determined by η_{out}. An estimate for the lower limit of the differential efficiency for short-cavity lasers can be obtained by assuming that for high threshold current densities the quantity η_{out} also remains constant and approximately equal to 0.6. This is apparently an underestimation since, as shown by calculations,[7] the absorption by free carriers in the active region and waveguide grows slower with decreasing laser cavity length than α_{out} does. Nevertheless, as obvious from Fig. 2, most of the short-cavity diodes with $I_{th} \geq 3$ kA/cm^2 have a lower differential efficiency than the lowest limit calculated in this way. This suggests the presence of additional mechanisms apart from leakage capable of reducing the differential quantum efficiency in short-cavity lasers with a high threshold current density.

Fig. 3 compares experimental data and calculated curves relating to the dependence of differential quantum efficiency on cavity length for AlGaAs/GaAs lasers prepared from structures with active region thicknesses of 60, 100, and 200 Å.

A comparison of Figs. 3 and 2 permits one to conclude that although the decrease of differential efficiency for AlGaAs/GaAs lasers with active regions 60 and 100 Å thick occurs only for the laser diodes with cavities two–tree times shorter than those of the InGaAsP/GaAs diodes, however as for the threshold current densities at which η_d is observed to drop, their values for InGaAsP/GaAs and AlGaAs/GaAs devices are very close. For AlGaAs/GaAs diodes with a 200 Å-thick active region which have the lowest threshold current densities for short-cavity devices, one does not practically see any drop in differential efficiency. This experimental result is in accordance with the calculations by which the decrease of η_d in such diodes should not exceed $\sim 5\%$ for threshold current densities ≤ 10 kA/cm^2. However because of a large bandgap difference between the cladding and the waveguide, theory predicts only a very weak decrease of η_d for AlGaAs/GaAs diodes with active regions 60 and 100 Å thick as well, whereas the experimental values of η_d for AlGaAs/GaAs diodes decrease with increasing current density nearly in the same way as they do for InGaAsP/GaAs lasers. (Curves (A) and (B) in Fig. 3 represent values of η_e and η_d for 60 Å thick active region lasers, calculated in the same way as in Fig. 2.) This result is another argument for the existence of other mechanisms, besides the current leakage included in the calculations, which provide a substantial effect on the falloff of differential efficiency. The filamentation of lasing could be one of possible causes for the drop of differential efficiency with decreasing cavity length in short-cavity laser diodes. Earlier studies[9] of the spatial distribution of near-field intensity for both types of the diodes under consideration revealed a dramatic degradation of near-field uniformity observed with decreasing cavity length.

The effect of excess current through the non-lasing regions of a stripe is seen in studies of the intensity of spontaneous emission from the active region beyond the lasing threshold (Figs. 6–7). In short-cavity diodes, spontaneous emission from the active region, as a rule, continues to increase above the lasing threshold, as this is shown in Fig. 7. At the same time for long-cavity diodes one observes typically the relations presented in Figs. 6 and 7 for the AlGaAs/GaAs and InGaAsP/GaAs devices, respectively. The constant level of emission intensity from the active region above the lasing threshold corresponds to the assumption of Fermi level pinning in the active region above the lasing threshold which was adopted in the calculations. Note that despite the Fermi level pinning the waveguide emission band intensity continues to grow also above the lasing threshold which is in satisfactory agreement with the calculations illustrated by solid lines in Figs. 6 and 7.

Summing up the results obtained in this work one can draw the following conclusions:
(*i*) The model based on the assumptions of quasineutrality and continuity of quasi Fermi levels at interfaces yields a correct description of the contribution of carrier recombination in the waveguide of SQW SCH lasers. Carrier leakage from the active region should be the major cause of the increase of the threshold current density in the QW lasers in question in the range of threshold current densities above 3 kA/cm^2 and bandgap difference between the cladding and waveguide layers of ≤ 100 meV.
(*ii*) It has been established that the non-equilibrium carrier concentration in the waveguide should continue to grow above the lasing threshold, which may cause an increase of leakage to the cladding with increasing threshold current density, and an anomalous decrease of differential efficiency in short-cavity diodes.

(*iii*) The experimentally observed decrease of differential efficiency in short-cavity diodes occurs substantially faster than it would follow from leakage calculations within the model used. An increase in lasing filamentation as one crosses over to short-cavity diodes could be one of the causes of the additional decrease in differential efficiency.

In conclusion, the authors express their gratitude to A.V. Kochergin for preparation of the InGaAsP/GaAs laser structures, and to Zh.I. Alferov for stimulating interest to the work.

References

[1] Zh.I. Alferov and D.Z. Garbuzov 1986 *18th Int. Conf. on Physics of Semicond., Stockholm (Sweden)* p. 203
[2] Zh.I. Alferov, D.Z. Garbuzov, S.N. Zhigulin, I.A. Kuz'min, B.B. Orlov, M.A. Sinitsyn, N.A. Strugov, V.E. Tokranov and B.S. Yavich 1988 *Sov. Phys.-Semicond.* **22(12)** 1334
[3] J.Z. Wilcox, S. Ou, J.J. Yang, M. Jansen and G.L. Peterson 1989 *Appl. Phys. Lett.* **55** 825
[4] S.R. Chinn, P.S. Zory and A.R. Reisinger 1988 *IEEE J. Quantum Electron.* **QE-24** 2191
[5] S.R. Chinn, P.S. Zory and A.R. Reisinger 1989 *SPIE* **143** 157
[6] Ya. Arakawa, M. Nishioko, N. Miura 1984 *Appl. Phys. Lett.* **45(1)** 7
[7] D.Z. Garbuzov, A.V. Ovchinnikov, N.A. Pikhtin, Z.N. Sokolova, I.S. Tarasov and V.B. Khalfin 1991 *Fiz. Tekh. Poluprovodn.* **25(5)** 928
[8] G.W.'t Hooft, C. van Opdorp and A.T. Vink 1983 *Acta Electronica* **25** 193
[9] D.Z. Garbuzov, A.V. Kochergin and E.U. Rafailov *5th All-Union Conf. on Physical Processes in Semicond. Heterostructures. Kaluga (USSR) 1990 (abstracts*, v. II) p. 68

Nonlinear effects in picosecond high-power diode lasers

E.L. Portnoi, E.A. Avrutin, and A.V. Chelnokov

A.F. Ioffe Physico-Technical Institute, Academy of Sciences of the USSR,
26 Polytekhnicheskaya st. 194021 Leningrad, USSR

Abstract. Dynamic properties of semiconductor laser diodes with fast intracavity saturable absorber are studied theoretically and experimentally. It is shown that passive Q-switching in such lasers produces picosecond optical pulses of very high intensity. Thus, gain nonlinearities greatly affect the properties of the lasers, leading to considerable reduction of self-pulsing current range, maximum optical pulse intensity, and repetition frequency obtainable. This effect is most pronounced for the shortest absorber relaxation times. Transient laser emission pictures observed experimentally are in reasonable agreement with the simulated results.

1. Introduction

The present paper continues the work on properties of semiconductor lasers with saturable absorber produced by deep implantation of heavy ions into the facet(s) of a laser diode.[1-5] In such an absorber, deep absorption modulation may be obtained along with very high recovery rates. The saturable absorber of this type was first proposed in Ref. 1 and utilizes the idea of spatial separation of charge carrier generation and recombination regions. This may be obtained by means of creating local amorphized regions with very high carrier recombination rates in the bulk of high-quality crystalline material. If the amorpized region dimensions are much smaller than the distance between these regions, then the total bulk fraction of the amorphized material is small. Therefore, the effective lifetime of nonequilibrium carriers may be drastically reduced while preserving such an important feature of pure material as the steep edge of the absorption spectrum. Hence, deep absorption modulation is possible due to dynamic Burstein–Moss (band filling) effect, as in normal direct-gap semiconductor. However, the recovery time of this effect is now defined by the rate of carrier capture by the amorphized regions and may thus be several orders of magnitude faster than the interband recombination time in the source material. Estimates made in terms of a model treating the amorphized regions as cylinders localized along the tracks of implantation ions predict picosecond relaxation times at actual implantation doses.[6] This is consistent with direct measurements of the lifetimes of nonequilibrium electrons in GaAs irradiated by oxygen ions.[7]

Stripe heterostructure laser diodes (LD's) with a saturable absorber of this kind display Q-switching regime producing optical pulses less than 10 ps in duration and more than 1 W in power. As the effective cross section of the laser waveguide is less than 10^{-7} cm^2, very high optical power density may be obtained inside it. Thus, gain nonlinearities are likely to affect the dynamic properties of the laser considerably.

2. Rate-equations analysis

As was often done previously,[8-11] we treat the laser under consideration as consisting of two sections. The first one is the gain section which in our case is the part of the LD active layer undamaged by the implantation, and the second is the saturable absorber section (implanted region) characterized by fast carrier recombination time τ_2. Within each section carrier density is supposed to be spatially uniform. Then, we may use the well-known system of rate equations for double-section LD[8-11] modified for our purposes as:

$$dn_1/dt = \eta_1 j/ed - n_1/\tau_1(n_1) - v_g g_1(n_1, S)S \qquad (1)$$

$$dn_2/dt = \eta_2 j/ed - n_2/\tau_2 + v_g \alpha_2(n_2)S \qquad (2)$$

$$dS/dt = \beta_{sp} B n_1^2 + [\Gamma v_g(r_1 g_1 - r_2 \alpha_2) - 1/\tau_{ph}]S \qquad (3)$$

$$1/\tau_1(n_1) = \tau_1^{nr} + Bn_1; \qquad \alpha_2(n_2) = a_2(N_{o2} - n_2)$$

$$g_1(n_1, S) = a_1(n_1 - N_{01})/(1 + \epsilon S) \qquad (4a\text{-}4c)$$

Here, three dynamic variables are carrier densities in the gain (n_1) and absorber (n_2) sections of the active layer, and the photon density S, assumed to be the same in both sections. For complete list of variables see Table. It also shows the default values, if any, for the parameters.

Rate equations in the form similar to (1–4) were previously used in quite a number of papers (see, e.g.[8-11] and the references therein). We believe, however, that the present model is the first to incorporate the gain compression term $1/(1+\epsilon S)$ into the analysis of double-section LD's. Gain compression factor ϵ is well known as an essential feature of any realistic model of dynamic (relaxation and modulation) properties of a single-section LD, that is, the LD without a saturable absorber. Different physical mechanisms have been proposed for gain compression, such as spectral[12] and spatial[13-14] hole burning, dynamic carrier heating by laser light,[15-17] nonlinear dispersion[18], etc. The relative contributions of these effects into the value of ϵ depend on the laser material and structure and are not important within a phenomenological model like the present one. From the results of Ref. 19, where the effect of gain nonlinearities on laser dynamics was clearly demonstrated, a simple estimate for photon density at which this effect becomes substantial may be easily obtained in the form: $\epsilon S > \tau_{ph}/\tau_1 \sim 10^{-3}$. As mentioned in the Introduction, the Q-switching is known to produce high light intensities, and hence this mild condition is likely to hold within a wide range of parameters. Therefore, we may expect the effect of nonlinearities to be considerable.

As the first step in the analysis of the system (1–4), one has to analyze its stationary solution, setting all the time derivatives $d/dt = 0$. This procedure leads, through simple computations, to a rather voluminous nonlinear arithmetic equation, which is a generalization of the cubic equation obtained in Ref. 11 in the limit of $\epsilon, \beta = 0$. Numerical solution for this equation gives the theoretical static light-current relationship for the laser. The result is presented in Fig. 1. The values of photon density are normalized to a natural scale value $S_1 = [v_g a_1 \tau_1(n_1^{th})]^{-1}$ which is the saturation photon density for gain section. As can be seen, the curve computed for the set of parameters typical for a LD with fast saturable absorber (the value $\tau_2 = 20$ ps was used) does not contain

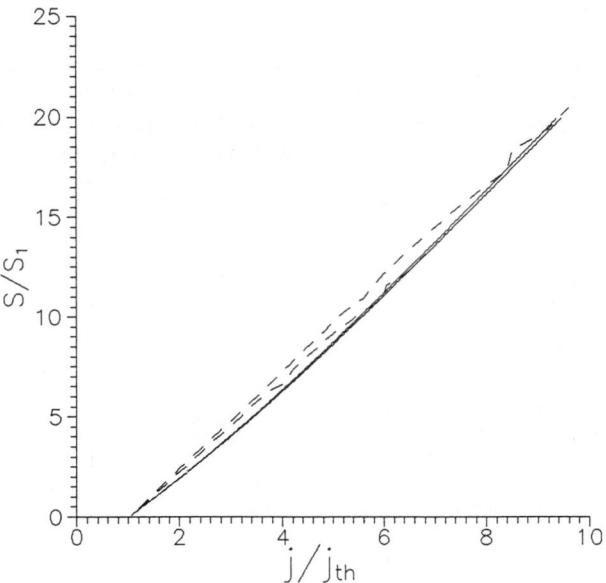

Fig. 1. Theoretical light-current stationary (solid curves) and time-avaraged (dashed curves) dependences for a LD with a saturable absorber. $\epsilon = 5 \cdot 10^{-18}$ cm^3 for the upper dashed curve; $\epsilon = 1 \cdot 10^{-17}$ cm^3 for the lower dashed curve. Effect of ϵ on the shape of the solid curve is negligibly small.

any sharp peculiarities, such as S-like region, and changes only slightly when gain compression factor is included. This is similar to the model predictions for the simple LD's without a saturable absorber, when the static properties of the laser are typically only slightly affected by gain nonlinearities.

The analysis of the self-pulsing, or Q-switching which is the same, regime proceeds with the small-signal (in)stability analysis of the stationary solution discussed hereabove. As the procedure had been previously discussed thoroughly in several papers (Refs. 9–11), only a short account is given here. Namely, three dynamic variables in (1–3) are allowed to possess small time-dependent deviations from the equilibrium (stationary) point. The system (1–4) is then linearized around this point. This results in a system of three linear differential equations for time-dependent deviations, the solution being sought in the form of $\sim \exp(\gamma t)$. Solution of the corresponding (cubic) secular equation for γ gives three eigenvalues. One of them (say, γ_1) is always real, the two others ($\gamma_{2,3} >$) may be, in general, either real or complex ($\gamma_2 = \gamma_3^*$). The latter was typically the case in our analysis. The condition Max$(\gamma_1, Re\gamma_{2,3}) > 0$ defines the range of system parameters over which the equilibrium solution is unstable, and the actual limit cycle of the system is an oscillatory process corresponding to self- pulsing (Q-switching) laser emission. Fig. 2 shows schematically the borders of this range in τ_2, j coordinates; all other parameters in (1–4) are fixed by the values of Table. Outside the self-pulsing range, stable laser emission is expected (bistable laser behavior is also possible[10] but it occurs for very slow saturable absorbers only and is not regarded here). A similar diagram of regimes had been previously obtained by Kawaguchi,[10] but only for $\epsilon = 0$ case. Besides, in Ref. 10 only the range of relatively long ($\tau_2 > 100$ ps) absorber relaxation times was regarded. Therefore, the range of pump currents enabling Q-switching was increasing monotonously with decreasing τ_2. The curves in Fig. 2 display somewhat more complicated behavior. As τ_2 is decreased down to a certain value τ_2', the current range for self-pulsing reaches a maximum, then its upper limit begins to fall and the lower one to rise. The rates of these dependencies are higher, the less is the value of ϵ. At a certain (small) critical value $\tau_2 = \tau_2''$ the equilibrium solution becomes stable at all values of pumping current. This result is quite clear qualitatively if one remembers that decreasing τ_2 not only makes the saturable absorber to respond faster to light changes but also increases the characteristic light intensity $S_2 = (v_g a_2 \tau_2)^{-1}$ required for the absorber to saturate. The first effect, promoting the self-pulsing behavior, is important at $\tau_2 > \tau_2'$ while the second one, suppressing the self-sustained pulsations, becomes dominating at smaller ($\tau_2 < \tau_2'$) values of recovery time. Kuznetzov[8] had introduced a simple criterion for self-pulsing (at $\epsilon = 0$) which with the notations of the present paper may be written as

$$p = a_2^2/a_1 \Gamma r_2 (N_{02} - \eta_2 \tau_2 j/ed) \tau_2 > 1 \qquad (5a)$$

To the first order in ϵ, this condition may be approximately generalized as

$$p' = p - \epsilon/(v_g a_1 \tau_{ph}) > 1 \qquad (5b)$$

With the realistic implantation-induced saturable absorber, one may assume, without any considerable loss of generality, that $\eta_2 = 0$ (ion implantation produces high serial resistance). Thus, p is directly proportional to τ_2, and (5) may be regarded as a condition to estimate τ_2''. The values of the parameter p corresponding to the values of τ_2'' obtained from the numerical analysis are of the order of unity ranging with ϵ from 0.9 to approximately 4. This is in reasonable agreement with the approximate analytical formula (5b).

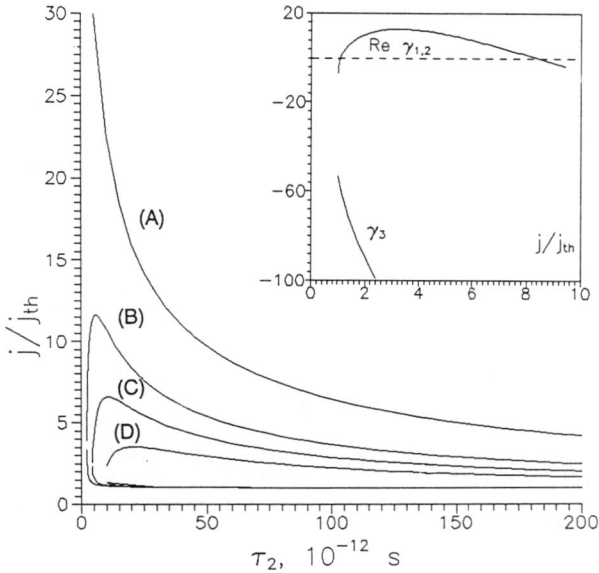

Fig. 2. Borders of the self-pulsing range for a LD with fast saturable absorber. (A) $\epsilon = 1 \cdot 10^{-19}$ cm^3, (B) $\epsilon = 5 \cdot 10^{-18}$ cm^3, (C) $\epsilon = 1 \cdot 10^{-17}$ cm^3, (D) $\epsilon = 2 \cdot 10^{-17}$ cm^3. The inset gives an illustration of the small-signal analysis (system eigenvalues versus normalized pump).

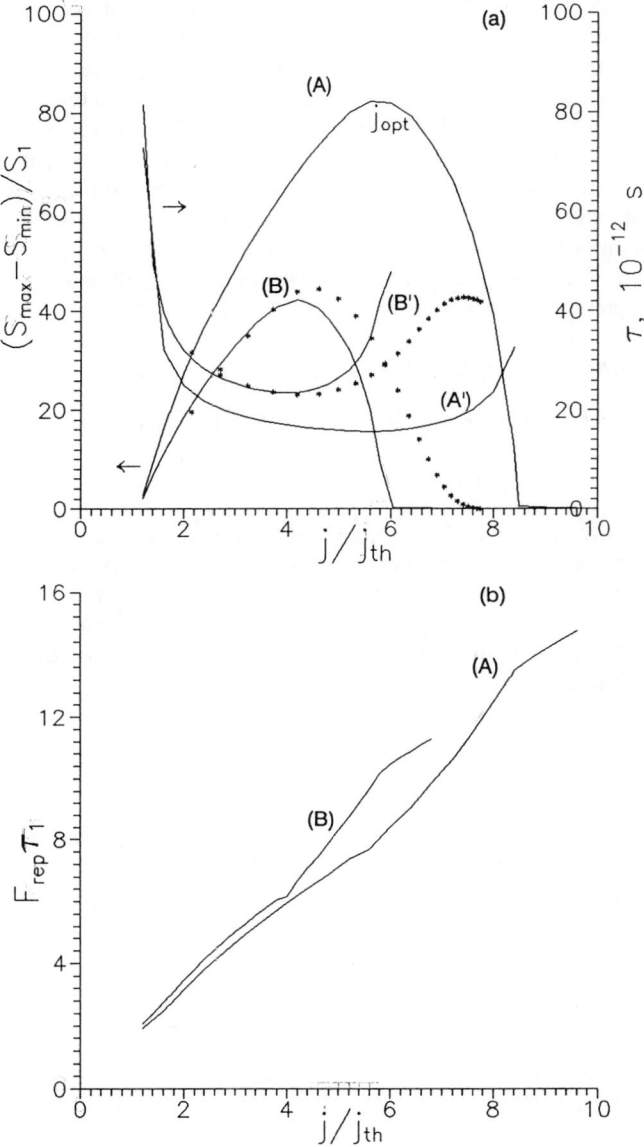

Fig. 3. Characteristics of limit-cycle pulses: (a) amplitude (A, B) and duration (A', B'); (b) repetition frequency $\tau_1(n_1^{th}) \approx 1$ ns. $\epsilon = 5 \cdot 10^{-18}$ cm^3 (curves A, A'); $\epsilon = 1 \cdot 10^{-17}$ cm^3 (curves B, B'). Asterisks show transient values at smooth pulse pump; $\epsilon = 1 \cdot 10^{-17}$ cm^3.

From (5b) it is clearly seen that the effect of nonlinearity on the self-pulsing regime is most pronounced at small τ_2 values. Indeed (see Fig. 2), including the gain compression term into the calculations results in the shift of τ'_2 point (that is, the value of absorber recovery time providing the widest pump current range for self-pulsing) towards larger values. For the set of parameters of

Table, τ_2' changes from a value of less then 1 ps at $\epsilon = 0$ to the figure of ≈ 6 ps at $\epsilon = 5 \cdot 10^{-18}$ cm^3 or ≈ 10 ps at $\epsilon = 1 \cdot 10^{-17}$ cm^3 therefore making the limitations set by high implantation doses (too small τ_2) more actual in practice. In general, introduction of the gain compression coefficient of the value, say, $\epsilon = 5 \cdot 10^{-18}$ cm^3 appears approximately equivalent to the reduction of absorber-to-gain cross section ratio a_2/a_1 from 5 to 2.2. This certainly means considerable narrowing of the self-pulsing range. A qualitatively similar result had been previously reported in Ref. 20, where a more sophisticated but less general model was applied for a specific kind of gain nonlinearity.

For further information, the system (1–4) was integrated numerically until the limit cycle for each set of parameters was reached. The results are shown in Fig. 3. Fig. 3(a) shows the values of pulse amplitudes and durations versus pump current; the corresponding values of pulse repetition frequency are plotted in Fig. 3(b). It is clearly seen that, in contrast to the predictions of a simplest analytical model,[8] the pulse amplitude and duration are more or less independent of the pumping value only in the narrow range of pump current values around the optimum value J_{opt}. As the pumping approaches either upper or lower limit of the self-pulsing range, the pulse amplitude smoothly falls down to zero. This is consistent with the small-signal analysis which has shown (see the inset in Fig. 2), that it is always the real part of the complex eigenvalue $\gamma_{2,3}$ that crosses zero and becomes positive. This is a characteristic feature[11] of smooth loss of stability, called also Hopf bifurcation. This general feature is, certainly, not affected by gain compression. However, including nonzero ϵ changes considerably the quantitative characteristics of the light pulsations, making them longer and weaker. Less sensitive to the value of ϵ, within the self-pulsing range, is the repetition frequency of the pulsations (Fig. 3(b)). However, as the upper limit of the self-pulsing range is reduced by gain compression, the maximum obtainable frequency is also reduced. Unlike the pulse characteristics, the time-averaged light-current dependence (dashed line in Fig. 1) is not practically affected by the nonlinearities, as well as by self-pulsing. This curve is fairly close to static light-current relationship plot, totally coinciding with it outside the self-pulsing current range.

Fig. 4 shows a characteristic simulated transient in the laser pumped by a smooth current pulse (dotted line) above j_{max} in amplitude. As can be seen, transient pump pulse amplitudes, as well as the "stationary" (limit-cycle) values shown in Fig. 3(a), depend nonmonotonously on pump. As the pump value approaches j_{max}, self-pulsing gradually fades down to smooth lasing. Note that the emission remains smooth even during the rear edge of the pump pulse where the current enters again the self-pulsing range, thus displaying a kind of hystheresis behaviour. To compare the "stationary" pulse parameters to the transient values, we have plotted the amplitudes and durations of Fig. 4 as asterisks in Fig. 3(a). As can be seen, transient optical pulse parameters under smooth pump pulse are fairly close to the stationary values at not too large currents. Only at pumping values near the cutoff value j_{max}, laser emission fails to follow the current quasistatically, thus keeping more pronounced pulsing than in "stationary" case.

In the analysis hereabove, we found it possible to keep only the gain compression term, neglecting similar nonlinearities for the saturable absorber. But as the relaxation time is less and the characteristic time constants of some of the nonlinearities are larger for the absorber than in case of the gain section, we believe that appropriate fitting of a_2 and τ_2 parameters may provide an adequate account for the saturable absorbtion compression. However, we have repeated the computations with the absorbtion compression included and found that it was of less importance than the gain compression as long as the coefficients ϵ for the gain and the absorber were kept of the same order of magnitude.

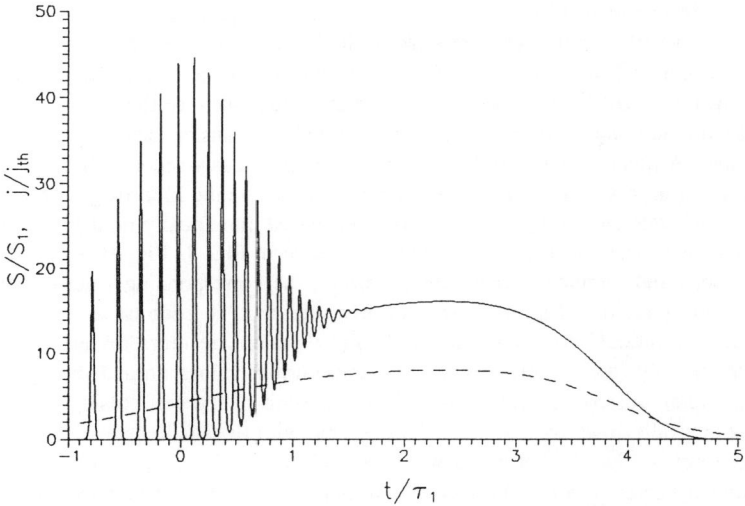

Fig. 4. Typical simulated transient laser emission under bell-shaped pulse pump. Pump pulse shape is shown by dashed line. $\epsilon = 1 \cdot 10^{-17}\,\mathrm{cm}^3$.

3. Experimental results and discussion

Fig. 5 shows the measured transient light emission from the AlGaAs/GaAs laser with the intracavity saturable absorber produced by deep implantation of oxygen ions.[4] The laser was pumped by current pulses of bell-shaped form. As shown hereabove, smooth pump pulse form enables one to treat the time-dependent pumping as approximately quasistationary. The different pump values in Fig. 5 may thus be attributed to three current ranges of Fig. 3: $j_{\min} < j < j_{opt}$; $j_{opt} < j < j_{\max}$; and $j > j_{\max}$. Indeed, the front edge of the plot in Fig. 5 demonstrates monotonous dependence of the optical pulse amplitude on current which is characteristic of small pumping values. Then, probably near the moment when the pump exceeds j_{opt}, an increase of pumping current leads to a decrease of the optical pulse amplitude. Note that this is less distinctly seen in the experimental picture of Fig. 5 than in the simulated one of Fig. 4, probably due to strong random fluctuations of pulse amplitudes. Finally, further increase of the pump switches the self-pulsing off. Irregular structure in the time-dependent lasing picture is due to the registration system resolution. Actually, some of the samples displayed passive modelocked behavior near the pumping pulse peak.[4] This proves that the value of τ_2 for these samples did not exceed the laser cavity round-trip time $\tau_{rt} = 2L/v_g \approx 5$ ps, $L = 200$ μm being the laser cavity length. Cease of Q-switching for such a laser may be considered as a proof, however indirect, for considerable effect of gain compression on the dynamic properties of the lasers studied. Indeed, the stability analysis reported hereabove predicts the self-pulsing for $\epsilon = 0$ at any reasonable current value provided that $\tau_2 \sim \tau_{rt}$, the absorber fraction is $r_2 = 0.1$ (experimental value), and the absorber-to-gain cross section ratio $a_2/a_1 > 3$ which seems to be realistic.

Certainly, the analysis within the framework of the simple rate equations is hardly applicable to the experimental situation directly due to both considerable uncertainty in parameters (such as a_2/a_1) and the oversimplifications of the model. The most important is the fact the rate equations essentially fail to describe laser dynamics on timescale less than (and probably close to) τ_{rt}.

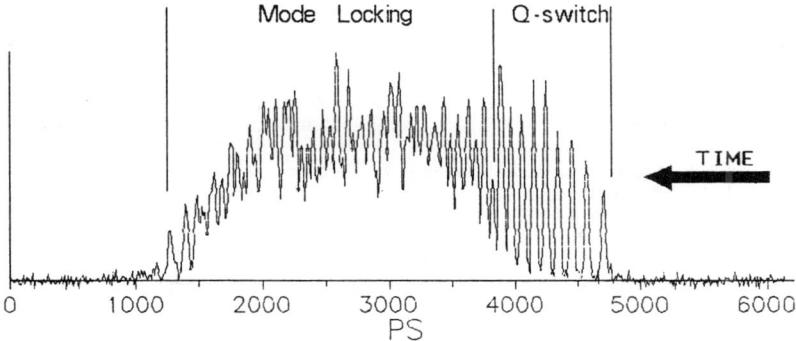

Fig. 5. Typical streak camera measured transient laser emission under bell-shaped pulse pump. Note the inverse (right-left) direction of the abscissa.

Therefore, predictions of the present model for the case of $\tau < \tau_{rt}$ should be treated with some care. Detailed character of transition from Q-switching to modelocking must be the subject for further study.

Tab. 1. List of the variables

Notation	Physical meaning	Value
n_1	Carrier density in gain section	
n_2	Carrier density in absorber section	
S	Photon density	
η_1	Pumping effectivity in gain section	1
η_2	Pumping effectivity in absorber section	0
j	Pump current	
e	Electronic charge	
$d.$	Active layer thickness	
τ_1	Carrier lifetime in gain section	
τ_2	Carrier lifetime in absorber section	
g_1	Optical gain coefficient	
α_2	Saturable absorption coefficient	
β_{sp}	Factor of spontaneous emission into lasing mode	$1 \cdot 10^{-4}$
Γ	Optical confinement factor	0.4
v_g	Group velocity of light	$0.75 \cdot 10^{10}$ cm^{-1}
r_1	Fraction of active layer occupied by gain section	0.9
r_2	Fraction of active layer occupied by absorber cection	$1 - r_1$
τ_{ph}	Photon lifetime due to unsaturable losses	1.8 ps
B	Spontaneous recombination coefficient	$3 \cdot 10^{10}$ m^3/s
τ_1^{nr}	Lifetime due to nonradiative recombination	10 ns
a_1	Gain cross section	$2 \cdot 10^{-16}$ cm^2
a_2	Saturable absorption cross section	$5a_1$
N_{01}	Effective carrier density for transparency	$1.2 \cdot 10^{18}$ cm^{-3}
N_{02}	Effective carrier density for transparency	$1.2 \cdot 10^{18}$ cm^{-3}
ϵ	Gain compression factor	

References

[1] Zh.I. Alferov, A.B. Zhuravlev, E.L. Portnoi, and N.M. Stel'makh 1986 *Sov. Tech. Phys. Lett.* **12** 452
[2] E.L. Portnoi, N.M. Stel'makh, and A.V. Chelnokov 1989 *Sov. Tech. Phys. Lett.* **15** 432
[3] E.L. Portnoi, N.M. Stel'makh, and A.V. Chelnokov,1989 *Sov. Tech. Phys. Lett.* **15** 865
[4] E.L. Portnoi and A.V. Chelnokov 1990 *Digest, 12th IEEE International Semiconductor Laser Conference (Davos, Switzerland)* 140
[5] N.M. Stel'makh, J.-M. Lourtioz, and F.H. Julien 1991 *Electron. Lett.* **27** 160
[6] E.A. Avrutin E.A. and M.E. Portnoi 1988 *Sov. Phys. Semicond.* **22** 968
[7] A.B. Zhuravlev, V.A. Marushchak, E.L. Portnoi, N.M. Stel'makh, and A.N. Titkov 1988 *Sov. Phys. Semicond.* **22** 217
[8] M. Kuznetzov 1985 *IEEE J. Quantum Electron.* **QE–21** 587
[9] L.A. Rivlin, A.N. Semenov, and S.D. Yakubovich *Dymamics and emission spectra of semiconductor lasers (Moscow, 1983)* (in Russian)
[10] H. Kawaguchi 1987 *Opt. Quant. Electr.* **19** Special Issue, p.S1–S37
[12] G.P. Agrawal 1987 *IEEE J. Quantum Electron* **QE–23** 860
[13] T.P. Lee, C.A. Burrus, J.A. Copeland, A.D. Dentai, and D. Marcuse 1982 *IEEE J. Quantum Electron.* **QE–18** 1101
[14] E.A. Avrutin 1990 *Sov. Techn. Phys. Lett.* **16** 387
[15] M.P. Kesler and E.P. Ippen 1987 *Appl. Phys. Lett.* **51** 1765
[16] B.N. Gomatam and A.P. De Fonzo 1990 *IEEE J. Quantum Electron.* **QE-26** 1689
[17] V.D. Pishchalko and V.I. Tolstikhin 1990 *Sov. Phys. Semicond.* **24** 288
[18] D.R. Hjelme and A.R. Michelson 1989 *IEEE J. Quantum Electron.* **QE–25** 1625
[19] C.B. Su and V.A. Lanciera 1986 *IEEE J. Quantum Electron.* **QE–22** 1568
[20] A.G. Plyavenek 1990 *Sov. J. Quantum Electron.* **3**

High frequency modulation of quantum well heterostructure diode lasers by carrier heating in microwave electric field

S.A. Gurevich
A.F. Ioffe Physico-Technical Institute, Academy of Sciences of the USSR,
26 Polytekhnicheskaya st. 194021 Leningrad, USSR

I.I. Filatov, B.M. Gorbovitsky, and V.B. Gorfinkel
Saratov Branch of IRE Academy of Sciences of the USSR,
21 Sakko i Vanzetty st. 410120 Saratov, USSR

Abstract. In this paper a new method of effective high frequency modulation of a QW diode lasers is discussed. The method is based on optical gain modulation by free-carrier plasma heating in external microwave electric field. The response time of carrier temperature is shown to be limited by the energy relaxation time, $\approx$ 1 ps in GaAs. The excess temperature of the plasma is about 100 K in the electric field of $\sim$ 1 kV/cm. Due to this the efficient gain modulation can be produced up to the frequences about 100 GHz. The laser output modulation by carrier heating is essentialy a parametric process. In this case, the amplitude modulation response is numerically calculated for SCH SQW laser. It is shown that in the frequency range 20 $\div$ 50 GHz, the modulation by carrier heating has a considerable superiority with respect to direct modulation by pumping current. Abrupt step-like variations of heating electric field allow to form picosecond optical pulses.

The experiments were carried out for a new ridge-guide SQW laser structure having two additional contacts for plasma heating. The results show the possibility of high frequency modulation and short optical pulse formation by the applied heating electric field.

1. Introduction

High frequency modulation of semiconductor lasers is a problem of importance for optical communications and high data-rate systems. At present the direct modulation of semiconductor diode lasers by means of pumping current variations is the most widely used technique. However, the practical limit of direct modulation bandwidth is commonly accepted to be about 10 GHz. To achieve much higher modulation frequencies a new method has been proposed,[1,2] based on optical gain modulation produced by free-carrier plasma heating in microwave electric field. It has been shown, that due to this modulation technique the modulation bandwidth could be extended up to the frequences of about 50 $\div$ 100 GHz. However, the practical problem to solve was how to apply the microwave electric field to the free-carrier plasma in the laser active layer. There was a number of earlier attempts to modulate the laser output by placing the samples into a microwave waveguide.[3] Unfortunately, in such kind of experiments the microwave electromagnetic field caused primarily inefficient direct modulation rather than direct influence on the free-carrier plasma. This was a result of screening by the top layers and the substrate of the laser structure.

In this paper we consider a new single quntum well laser structure with two additional ohmic contacts used for plasma heating. The results of experiments with this laser structure show efficient modulation of the laser output. In addition, we present some new results of theoretical study of the heating field effect on static and dynamic properties of quantum well lasers.

2. Experimental results

The fabricated laser structure is shown schematically in Fig. 1(a). It is essentially a common ridge-guide structure. For the laser fabrication we used AlGaAs-GaAs separate-confinement heterostructure (SCH) single quantum well (SQW) wafer grown by MBE on a p^+-GaAs substrate. The thickness of the QW active layer was 100 Å. The thickness of undoped waveguide layers was 0.2 μm. The ridge-guide structure was formed by etching of two parallel grooves, each of 4 μm width, sep-

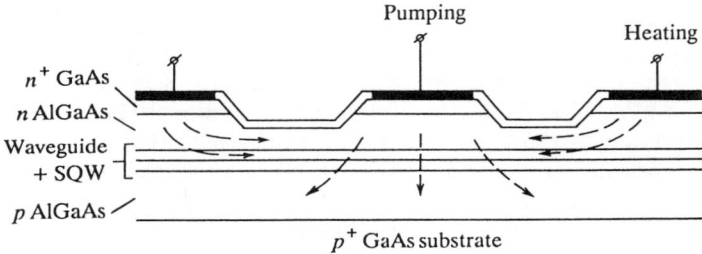

Fig. 1. (a) schematic AlGaAs-GaAs SCH SQW ridge-guide laser structure with two side contacts for plasma heating; (b) SEM photograph of the laser sample (top view). The gate-like contact area in the center is used for laser pumping.

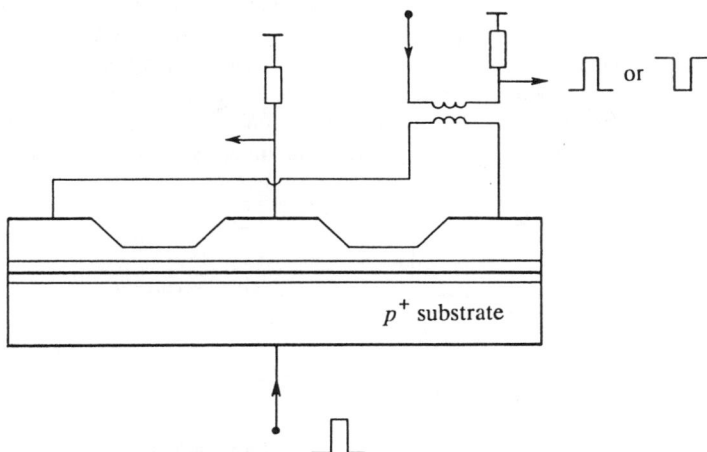

Fig. 2. The schematic diagram of the laser connection.

arated by 6 μm, which is the ridge width. Then, the surface of the grooves was covered by SiO_2 and ohmic contacts were embedded onto the ridge top as well as on the surface of the structure outside the grooves. An ohmic contact was made on the substrate surface too. A SEM photogaph of the laser sample is shown in Fig. 1(b). The central (gate-like) contact was used for laser pumping while the side contacts (source and drain) for carrier heating.

The schematic diagram of laser connection is shown in Fig. 2. To avoid excess ohmic heating we supplied the synchronized 100 ns current pulses with repetition rate of 40 kHz in both pumping and side-contact circuits. These circuits were separated by means of a high frequency transformer, as it is shown in Fig. 2.

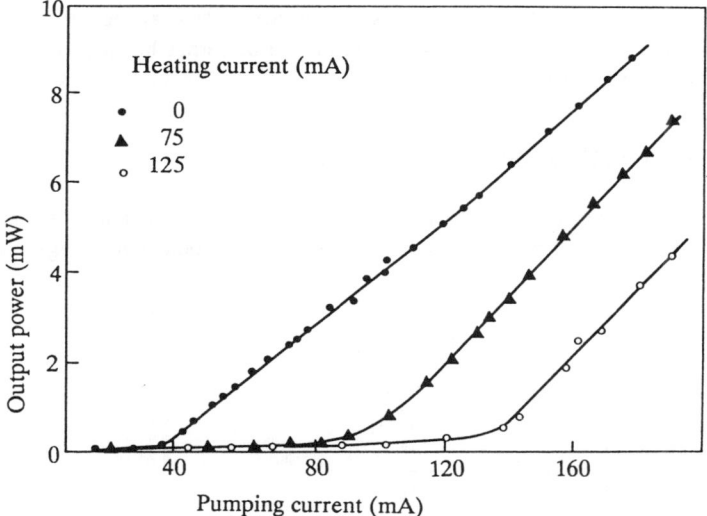

Fig. 3. Output power vs pumping current characteristics at different currents flowing through the side contacts.

The current-voltage characteristics measured at the side contacts were linear and with the laser cavity length of 400 μm the resistance was about 100 Ω at zero pumping and decreased with the increase of the pumping current. On the other hand, the pumping current was not practically influenced by the current flowing through the side contacts. Fig. 3 shows the set of output power vs pumping current characteristics obtained at different currents applied to the side contacts. The significant increase of the laser threshold manifests the influence of applied current on the optical gain of the laser.

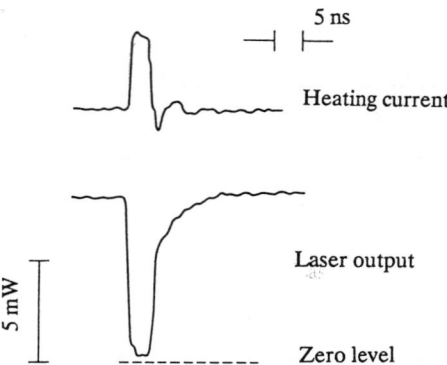

Fig. 4. Laser output response to a short heating current pulse.

The laser output response to the short 5 ns heating current pulse is shown in Fig. 4. The laser was initially driven above the threshold to the output power of 8 mW. Under the heating current pulse (150 mA peak value) the laser output was suppressed practically to zero. With the 14 μm gap between the side contacts, the estimated electric field applied to the free carrier plasma was about 6 kV/cm in this experiment. About 100% modulation efficiency of the laser output has been obtained also with 4 GHz alternating current signal when the laser was driven at twice the threshold. In our opinion, the results of these experiments show at least a new way of highly efficient modulation of laser output. Direct confirmation of carrier heating in the fabricated laser structure is the subject of further investigations.

3. Theoretical analysis and numerical results

In order to describe the effect of external electric field on steady-state and dynamic characteristics of the diode laser, we used a model based on laser rate equations and energy balance equation descibing the carrier heating

$$\frac{dn}{dt} = \frac{J}{ed} - \frac{n}{\tau_s} - \frac{c}{N}g(n,T)S, \tag{1}$$

$$\frac{dS}{dt} = \Gamma\frac{c}{N}g(n,T)S - \frac{S}{\tau_p} + \Gamma\frac{\beta n}{\tau_s}, \tag{2}$$

$$\frac{dT}{dt} = \frac{2}{3k}e\mu(n,T)E^2(t) - \frac{T-T_0}{\tau_\varepsilon(n,T)}, \tag{3}$$

where n and S are the carrier and photon densities, respectively, $E(t)$ is the applied electric field, $\mu(n,T)$ is the mobility, T is the temperature of electron-hole plasma, T_0 is the lattice temperature, and $\tau_\varepsilon(n,T)$ is the energy relaxation time. The other symbols have their usual meanings. The dependence of the optical gain on carrier concentration and electron-hole plasma temperature was calculated assuming Fermi distributions for both types of carriers and step-like density of states in the QW active layer. The energy (and polarisation) dependent matrix elements for the interband optical transitions were taken similar to those used elsewhere.[4] The energy relaxation time τ_ε and the momentum relaxation time τ_k determining the electron mobility have been calculated by Monte-Carlo technique.[5] In these calculations the carrier interaction with polar optical phonons was assumed to be the major scattering mechanism.

Generally, equations (1), (2), and (3) have to be solved self-consistently and this is the subject for numerical calculations. But in the frame of small-signal analysis, some useful results can be obtained in the analytical form.

In the steady-state case, we have calculated the dependence of laser threshold current on heating electric field applied in the direction parallel to the active layer plane. The results are shown in Fig. 5. In the calculations we used the parameters corresponding to the SCH SQW laser structure mentioned above. For the solid curves plotted in Fig. 5 the parameter is the low-field electron mobility. The dashed curve corresponds to the case when the hole heating was not regarded and the hole temperature was assumed to be equal to the lattice temperature (300 K). The main result illustrated by Fig. 5 is a sharp increase of the laser threshold current under the heating field ranging from 1 kV/cm to 6 kV/cm. In such fields, the excess temperature of the carrier is about $100 \div 300$ K, so that intervalley trasitions and thermal emission processes are not very important.[2]

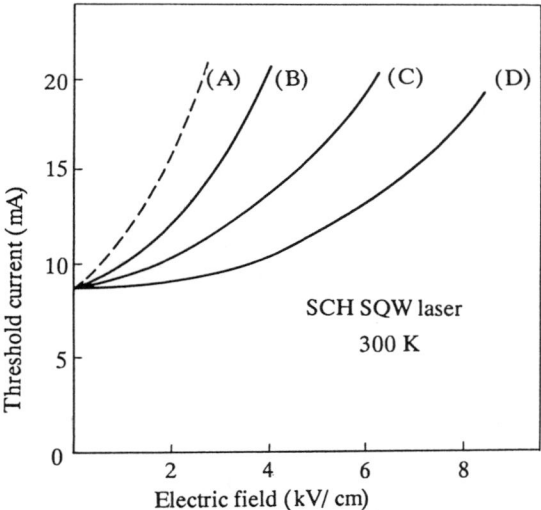

Fig. 5. Calculated dependence of AlGaAs-GaAs SCH SQW laser threshold current on the electric field. Low-field electron mobilities are: (A),(B) 8000 cm²/Vs, (C) 4000 cm²/Vs, (D) 1200 cm²/Vs. Dashed curve corresponds to the case when the hole temperature is set to be equal to the lattice temperature.

To characterize the dynamic behavior of a laser under the gain modulation by the free-carrier heating, we start with a small-signal approach to the equations (1)-(3). We consider the small-signal modulation by the electric field $E(t) = E_0 + \delta E \sin(\omega t)$ with alternating part amplitude

$\delta E \ll E_0$. As a result, one may expect small variations of all the variables, in particuliar, those of the carrier concentration $\delta n \ll n_0$, where n_0 is the steady-state concentration. With this, the mobility and energy relaxation time can be treated as constant parameters and equation (3) can be solved independently for the carrier temperature response δT. Finally, to linearize the rate equations (1) and (2) we expand the gain

$$g(n,T) = g(n_0, T_0) + \frac{\partial g}{\partial n}\delta n + \frac{\partial g}{\partial T}\delta T, \qquad (4)$$

where the differential gain term $\partial g/\partial T$ describes the gain modulation by carrier heating.

As one can see from (3), the characteristic time for the carrier temperature response should be as short as a few picoseconds due to the small value of $\tau_\varepsilon \approx 10^{-12}$ s. However, the laser output response δS which is of primary interest here, may be not so fast because it involves the carrier concentration variation controlled by the spontaneous lifetime $\tau_s \approx 10^{-9}$ s.

For the amplitude modulation response, the small-signal analysis gives

$$|\delta S(\omega)| = \frac{\tau_p}{\tau_s}\Gamma \frac{\partial g/\partial T}{\partial g/\partial n}\delta T A_T(\omega). \qquad (5)$$

The frequency dependent part $A_T(\omega)$ is

$$A_T(\omega) = \frac{\omega_0^2(1+\omega^2\tau_s^2)^{1/2}}{[(\omega_0^2-\omega^2)^2+\gamma^2\omega^2]^{1/2}}, \qquad (6)$$

where ω_0 is the electron-photon resonance frequency, γ is the damping factor, both depending on the output power and structure parameters.[7] For comparison, the frequency dependent part $A_J(\omega)$, corresponding to the direct modulation is given by

$$A_J(\omega) = \frac{\omega_0^2}{[(\omega_0^2-\omega^2)^2+\gamma^2\omega^2]^{1/2}}. \qquad (7)$$

It is easy to see that in the case of small damping the $A_T(\omega)$ and $A_J(\omega)$ functions have maxima near the frequency ω_0. The most important feature is that in the high frequency limit $A_T(\omega)$ is proportional to ω^{-1} while $A_J(\omega) \propto \omega^{-2}$. This strictly indicates that the modulation by carrier heating is very promising for high frequency operation.

For a quantitive example, we have calculated numerically the amplitude modulation response of a QW diode laser under the sisnusoidal variation of heating electric field and that of pumping current. The results are plotted in Fig. 6. As shown in Fig. 6, both curves have a maximum at the frequency $f = \omega/2\pi \approx 10$ GHz, which corresponds to the electron-photon resonance. In this case of modulation by electric field, this resonance is more pronounced. It is important to note that in the frequency range $20 \div 50$ GHz, which is above the resonance, the modulation by carrier heating has a considerable superiority with respect to direct modulation.

In the quantitative example given above the alternating component of electric field δE and that of current δI were taken to be 5% of their constant components. Nevertheless, the conditions of small-signal response were actually fulfilled in the low-frequency and high frequency ranges, but not in the vicinity of the resonance. This is why in this case we solved the equations (1)–(3) numerically instead of using the formulas (6) and (7).

In contrast to direct modulation, the considered modulation of the optical gain by carrier heating is essentially a parametric process. Its parametric nature results in a quite different amplitude response of a laser (Fig. 6). Moreover, the effects of frequency mixing, harmonics excitation and period doubling can be observed in this case. The important feature of the laser output response to the alternating heating electric field is that the result strongly depends on the ratio of constant to alternating components of the applied field. For instance, if the constant component is zero, the modulation at the frequency f will lead to the output response at the frequency $2f$. This frequency doubling is due to the simple fact that the effect of carrier heating does not depend on the direction of field vector in the active layer plane.

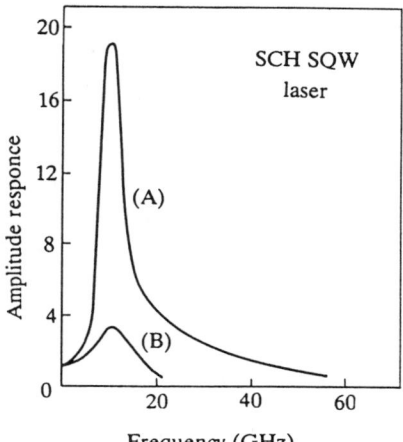

Fig. 6. Calculated amplitude modulation responses of SCH SQW laser. (A) modulation by carrier heating in alternating electric field $E(t) = E_0 + \delta E \sin(2\pi ft)$; $E_0 = 8.4$ kV/cm, $\delta E = 0.05 E_0$. Constant pumping current $I_0 = 3 I_{th}$. (B) direct modulation by pumping current $I(t) = I_0 + \delta I \sin(2\pi ft)$; $I_0 = 3\ I_{th}$, $\delta I = 0.05\ I_0$. Electric field is not applied: $E(t) = 0$.

In order to evaluate the potential of gain-modulated lasers in high bit-rate systems, we have carried out a numerical simulation of the laser reaction to a step-like variation of the heating electric field. If the laser is initially driven above the threshold, the reaction to the abrupt step-like increase of the applied electric field is a fast decrease of the output. In this case, the characteristic transient time is a few picoseconds determined by the carrier temperature rise and the photon lifetime in the resonator. Depending on the amplitude of the electric field step, the laser will be either switched off or brought into relaxation oscillations after the fast drop of the output. Somewhat more complicated is the laser output reaction to a step-like decrease of the heating field (Fig. 7). If the laser is initially above the threshold, the sequence of events will occur as follows. As a direct reaction to the switching off the heating field, the carrier temperature will drop during a few picoseconds. Due to this drop, the optical gain will rise initiating a fast increase of the output power. But the increasing photon density will start to consume the inversion, and hence the optical gain will start to drop. This process forms the back front of the optical pulse. The analysis shows that the duration of optical pulse generated in this process is determined by the ratio of the steady-state inversion concentration before the electric field step to that after the step. The example given in Fig. 7 demonstrates a 10 ps optical pulse.

In the light of the present results the best way for high bit-rate modulation is probably the superposition of short negative pulses against a positive background of the heating field. Indeed, after such negative pulse the return of the laser under the action of heating field will suppress the relaxation oscillations being source of bit error. The ability of the laser to respond to very high frequency periodic modulation by the electric field is promising for high-capacity optical communication systems. Using the simultaneous modulation by high frequency heating field and by pumping current at lower frequences one may separate a number of channels by their carrying frequency. In this case all the advantages of the frequency multiplexing system can be employed.

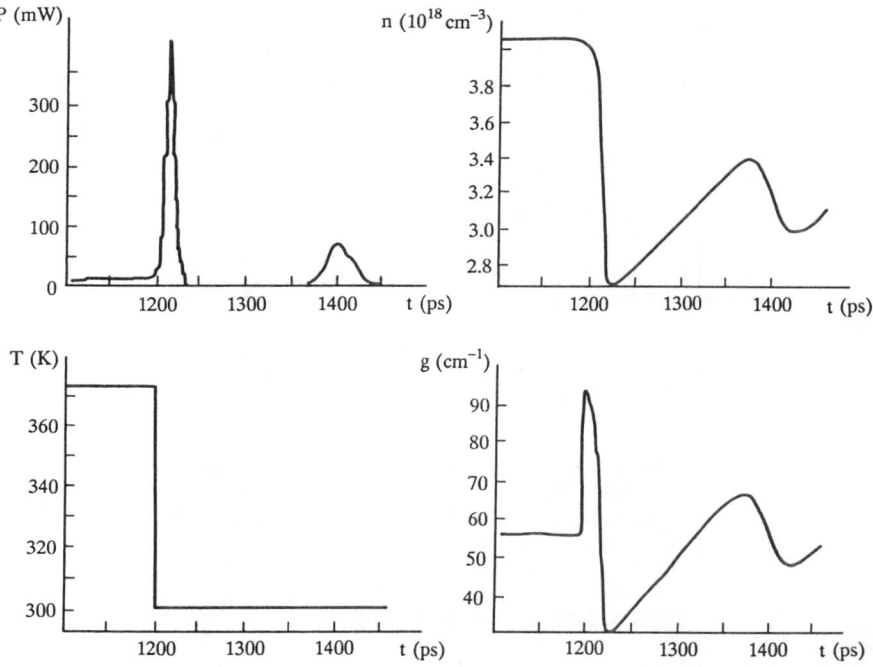

Fig. 7. The reaction of laser output, electron temperature, gain and carrier concentration on the abrupt step-like decrease of heating electric field. The field amplitude is $E_0 = 8.4$ kV/cm; the pumping current $I = 1.5\, I_{th}$.

To summarize, we have shown that carrier heating in external electric field is very promising for modulation of laser diode output up to the frequencies of $50 \div 100$ GHz. A new laser structure has been designed to realize effective modulation by carrier heating.

References

[1] V.B. Gorfinkel and I.I. Filatov 1990 *Sov. Phys.-Semicond.* **24** 4 466
[2] V.B. Gorfinkel, B.M. Gorbovitsky, and I.I. Filatov 1990 *Int. J. of Infrared and Millimeter Waves* (to be published)
[3] S. Takamija, F. Kitasawa, and J.I. Nishizawa 1968 *Proc. IEEE* **56** 1 135
[4] D. Ahn and S. Chung 1990 *IEEE J. Quantum Electron.* **QE-26** 1 13
[5] V.B. Gorfinkel and S.G. Shafman 1988 *Sov. Phys.-Semicond.* **22** 5 500
[6] K.Y. Lau and A. Yariv 1985 *Semiconductor and Semimetals* **22** Part B 79

Active Q-switching of GaAlAs/GaAs lasers using free carrier effects in a modulation doped QW

Yu.G. Kozlov, T.V. Shubina, A.A. Toropov, and I.Yu. Shvechikov

A.F. Ioffe Physico-Technical Institute, Academy of Sciences of the USSR,
26 Polytekhnicheskaya st. 194021 Leningrad, USSR

Abstract. Active Q-switching was studied in a two-segment AlGaAs/GaAs DH laser with a modulation-doped GaAs QW placed in the n-region of a p-n junction. The modulator uses the blue shift of the QW absorption edge due to band filling by a 2D electron gas. The electron gas concentration could be varied in the range $0 \div 8 \times 10^{11}$ cm^2 by applying bias voltage. Only about 100 mV of the modulating voltage was necessary to provide stable active Q-switching so that every period of the modulating microwave signal gave rise to one optical laser pulse less than 50 ps FWHM. The threshold injection current density was very low and ranged from 400 to 800 A/cm^2. The modulation bandwidth has been estimated to be not less than $4 \div 5$ GHz.

1. Introduction

Active Q-switching is a very attractive method to control laser generation and to produce ultrashort optical pulses because lower modulation power is required than, for example, for gain switching. This is usually attained in a two-segment laser consisting of an optical amplifier and an electroabsorption loss modulator. The amplifier is driven by injection current while the modulator uses of some electrooptical effect and is controlled by an external bias voltage. Different electrooptical phenomena have been utilized to provide convenient operation of the devices. These are the Franz–Keldysh effect in a double heterostructure[1] and the quantum confined Stark Effect (QCSE) in a multiquantum well laser structure.[2] There is one common problem in utilizing these effects. There is noticeable absorption even in the most transparent state of the modulator at the wavelength corresponding to the generation of a modulator-free laser. This seems to be the reason for increased threshold current of the two-segment laser.[3] An increase in the electric field applied to the modulator active region results in additional enhancement of absorption below the band gap. This occurs due to both the shape alteration of the band edge caused by the Franz–Keldysh effect and the red shift of the excitonic peak due to the QCSE.

To solve this problem, we propose to use free carrier effects in a modulation-doped QW, in particular, the blue shift of the QW absorption edge due to the band filling by 2D electron gas.[4,5] The latter is very similar to the Moss–Burstein effect in doped bulk semiconductors. Free electrons present in an n-modulation-doped QW fill the lowest confined states up to the Fermi level. For 2D gas of electrons of mass m_e, the Fermi energy E_F is equal to $(\pi \hbar / m_e) \times N_s$ where N_s is the sheet concentration. In GaAs, for example, E_F is about 35 meV when $N_s = 1 \times 10^{12}$ cm^{-2}. Light absorption can be associated only with the transitions to unfilled electronic states. So, one may expect the absorption edge to be shifted towards higher energies than in an undoped QW and the energy shift $\Delta E = (1 + m_e/m_h)E_F$ is proportional to N_s. This allows in principal to reduce the modulator losses and, hence the injection current density of the two-segment laser. For this purpose we used an AlGaAs/GaAs *p-n* laser heterostructure with an undoped GaAs single QW placed near the *p-n* junction in the *n*-doped region. The advantage of this structure is that it allows to control the electron population in the QW by applying the bias voltage to the *p-n* junction. To elucidate the voltage dependence of N_s and electrooptical characteristics of the structure, spectral studies were made of the photocurrent and the waveguide light transmission. Efficient active Q-switching was obtained and studied in two-segment lasers fabricated from the structures.

2. Samples and spectral waveguide technique

The structures under study were grown by MBE and low-temperature LPE. The threshold injection current density ranged from 200 to 500 A/cm^2. The paper describes mainly an MBE structure as the others demonstrated a similar behaviour.

Fig. 1 shows a schematic view of the structure. An undoped GaAs QW of about 120 Å wide is put in the center of a 0.4 μm wide Al$_{0.28}$Ga$_{0.72}$As optical waveguide *n*-doped up to 1.3×10^{17} cm^3. The waveguide is surrounded by Al$_{0.48}$Ga$_{0.52}$As confining layers. An abrupt asymmetric *p-n* junction is formed at the interface between the waveguide and the *p*-doped upper confining layer.

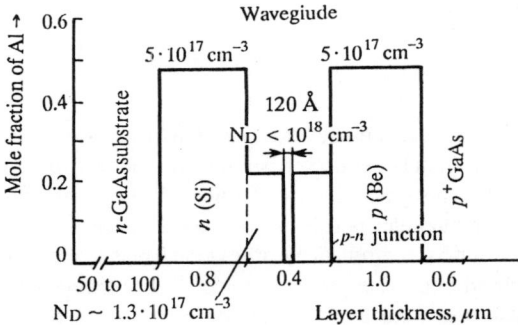

Fig. 1. Schematic view of a modulation-doped QW structure

The slab waveguide geometry of the samples was used to study the waveguide light transmission. The emission of an iodine tungsten lamp transmitted through a monochromator and polarizer was focused onto the slab guide sample placed between two gold contacts. The emission from the sample was collected with a microscope and sensed with a photomultiplier. The technique allowed to measure the spectra of absorption, electroabsorption, and photocurrent selectively for the TE and TM waveguide modes.

3. Electrooptical characteristics of QW waveguide structures

The 2D electron gas present in the QW is supposed to be responsible for the electrooptical characteristics, so it is important to know the dependence of its concentration N_s on the external bias. For this purpose the photocurrent was measured as a function of the reverse bias in a sample long enough to absorb practically all the photons above the QW absorption band edge. So, the photocurrent was almost independent of the absorption coefficient and reflected the extension of the space depletion regions (SDR's) in the structure.

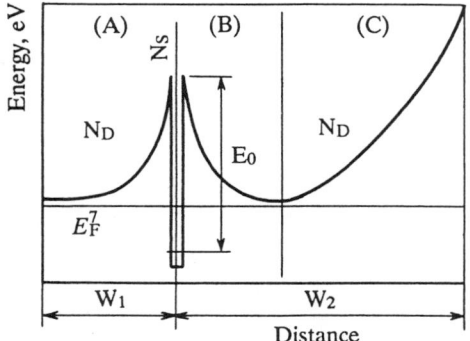

Figure 2. Schematic diagram of the potential and space depletion regions (A, B, C) in the structure. W_2 is the distance from the QW to the p-n junction interface. N_D is the n-region doping level.

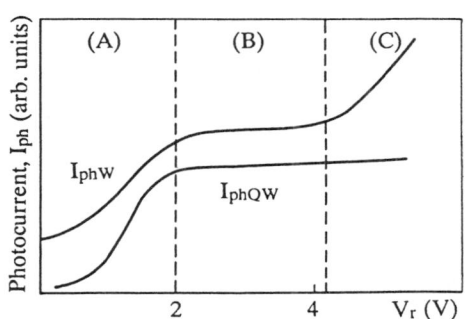

Figure 3. Photocurrent as a function of the reverse bias V_r: (I_{phQW}), photocurrent from the QW; (I_{phW}), photocurrent from the waveguide.

There were three depletion regions in the structure: one on each side of the QW and one more is the p-n junction SDR (see Fig. 2). The photocurrent was measured as a function of the reverse bias voltage at two different wavelengths. The first wavelength is chosen above the GaAs QW band edge but below the band edge of the AlGaAs waveguide. That provided the registration of the photocurrent from the QW only. The photocurrent (I_{phQW}) is plotted in Fig. 3 against the reverse bias voltage. It increases rapidly up to 2 V and saturates above this value. It seems likely that the reverse bias produces the extension of the p-n junction SDR which overlaps completely one of the two QW depletion regions at about 2 V. At this voltage almost all the photoexcited carriers are separated and its further increase can no more increase the photocurrent.

In the second measurement the light wavelength was chosen above the band gap of the AlGaAs waveguide, so it was a non-guided operation. In this case the photocurrent is the sum of the values from the QW and waveguide material in the SDR's. The voltage dependence of this total photocurrent (I_{phW}) is shown in Fig. 3. The curve has three portions. Below 2 V the photocurrent increases rapidly, which corresponds to the convergence of the SDR's. At higher voltage the QW photocurrent becomes saturated and further increase is due only to the enhancement of photocurrent from the waveguide SDR's. But the p-n junction SDR and one of the two QW SDR's are completely overlapped and the enhancement is proportional to the extension of the other QW SDR. Whilst there is 2D electron gas in the QW it screens the potential and prevents the extension of this region. Therefore, the photocurrent is almost voltage-independent until the full depletion of the QW occurs at about 4 V. After that the photocurrent varies with voltage as in a simple p-n junction.

To describe the details of this process, it is necessary to solve self-consistently the Poisson and Schrödinger equations. However, the QW width is small compared to the width of the SDR's, so one can neglect the QW shape alteration and the Stark shift of the confined levels. With this assumption the photocurrent data can be interpreted in the framework of a simple electrostatic model. The model also implies that the SDR's of the p-n junction and the QW are fully overlapped and the photocurrent increase is proportional to the extension of the other QW SDR (region (A) in Fig. 2). Besides, the electron Fermi level is supposed to be fixed throughout the sample. The Poisson equation solved with these assumptions gives the following expressions for the voltage V_0 of complete depletion of the QW, sheet electron concentration N_s and the width of the QW depletion region W_1:

$$V_0(E_0) = \frac{(W_2\sqrt{eN_D} + \sqrt{2E_0\varepsilon})^2}{2\varepsilon}. \tag{1}$$

For $V > V_0$ $N_s = 0$ and

$$W_1(V) = \sqrt{\frac{2\varepsilon V}{eN_D}} - W_2. \tag{2}$$

For $V < V_0$,

$$N_s(V) = \frac{N_D[W_1(V) + W_2]^2}{2W_2} - \frac{\varepsilon V}{eW_2} \tag{3}$$

and

$$W_1(V) = \frac{\sqrt{f(V, E_0)} - AeN_D W_2}{eN_D(\frac{e}{\varepsilon}W_2 + A)}, \tag{4}$$

where

$$f(V, E_0) = 2eN_D(\frac{e}{\varepsilon}W_2 + A)(E_0 W_2 e + A\varepsilon V) - \frac{e^3 W_2^3 N_D^2 A}{\varepsilon}. \tag{5}$$

Here e, ε, and V denote, respectively, the elementary charge, the permittivity and the full voltage ($V = V_r + V_{\text{built-in}}$); V_r is the reverse bias voltage; A is given by $\pi\hbar^2/m_e$. The parameters W_2, E_0, N_D are explained in Fig. 2.

Assuming the photocurrent to be proportional to $W_1(V)$ one can express its voltage dependence as

$$I_{ph}(V) = \frac{C}{\sqrt{N_D}}\sqrt{\frac{2\varepsilon}{e}}(\sqrt{V} - \sqrt{V_0}) \tag{6}$$

for $V > V_0$, and for $V < V_0$

$$I_{ph}(V) = \frac{C}{\sqrt{N_D}} \times \frac{\sqrt{f(V)} - \sqrt{f(V_0)}}{e(\frac{e}{\varepsilon}W_2 + A)}. \tag{7}$$

Here C is an adjustable parameter.

These dependences, joined together at the point $V = V_0$, are plotted in Fig. 4. They show good agreement with the experimental data, which proves the validity of the simplified description. The calculated voltage dependence of N_s is also shown in Fig. 4. The increase in the bias from 2 to 4.4 V decreases N_s from 8×10^{11} cm^{-2} to zero.

The high value of N_s predicts a significant blue shift of the absorption edge associated with the optical transitions from the lowest hole states to the lowest electron level in the QW due to the band filling. To detect this phenomenon the spectra of the waveguide absorption for the TE and TM waveguide modes were measured at different bias voltages at room temperature.

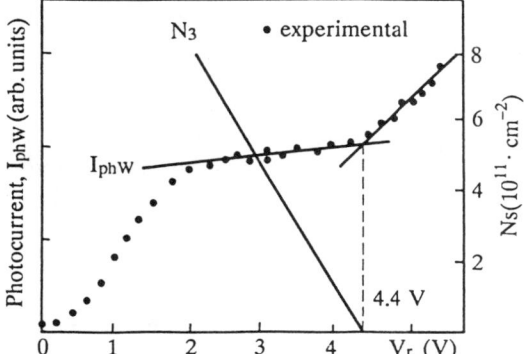

Fig. 4. Photocurrent I_{phW} and 2D electron gas concentration N_s as a function of the reverse bias V_r. Circles: experimental points. Solid line: calculation.

The spectra in Fig. 5 demonstrate well-pronounced absorption peaks. An increase in the reverse bias changes significantly the band edge shape. First of all, there is a noticeable enhancement of absorption in the whole spectral region of the band edge. Then there is an insignificant shift of the peaks. Finally, we observe the appearance and a large red shift of a specific "wave" of absorption above 2 V of reverse the bias. The last feature can be clearly seen in the differential electromodulation spectra of transmission measured at different constant voltages (see Fig. 6). The specific resonance-shape differential spectrum, which includes regions of increasing and decreasing absorption (positive and negative signal), appears at about 2 V, shifts approximately by 50 meV without significant increase in the spectral width and then disappears gradually above 4 V.

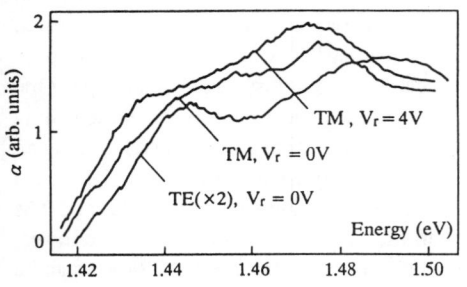

Figure 5. Absorption spectra of the TE and TM waveguide modes.

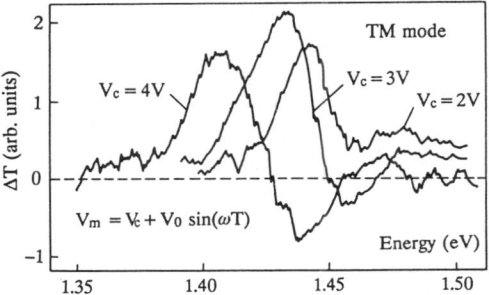

Figure 6. Differential electromodulation transmission spectra of the TM mode.

Fig. 7 shows the energy position of the spectral singularities as a function of the bias. The characteristic energy of the "rapidly" shifted resonance is determined as a zero of the electromodulation spectra. The singularities can be divided into two types according to their dependence on the voltage. There are "slowly" shifted absorption peaks for the TE and TM waveguide modes and the "rapidly" shifted resonance observed in both polarizations. The total shift is about 6 meV for the "slowly" shifted peaks and about 50 meV for the "rapidly" shifted resonance.

It is very likely that the lowest spectral singularity corresponds to the lowest possible optical transition in the QW. We mean the transition between the lowest hole and electron states, when there is no electron gas in the QW, and the transition from the lowest hole state to above the Fermi edge when there is 2D electron gas of a concentration N_s in the QW. One can check this assumption comparing the slope of the experimental dependence with that estimated for the Moss–Burstein shift, using the data of the photocurrent measurements. The calculated line plotted in Fig. 7 is adjusted to the experimental one at the point of the lowest voltage. The agreement is rather good at high N_s, but there is some divergence at low electron concentration.

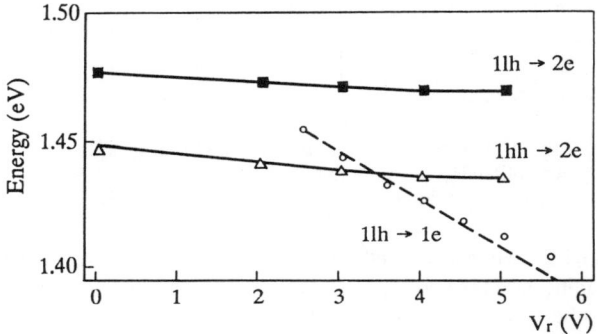

Fig. 7. Energy position of the spectral singularities against the reverse bias. Position of the "slowly" shifted peaks of the TE (triangles) and TM (squares) waveguide modes. Position of the "rapidly" shifted singularity of the TM mode absorption (full circles). Broken line—the computation for the Moss–Burstein shift based on the photocurrent data (the slope $E/V_r = 18.3$ meV/V). (See text).

There is a number of other electrooptical effects in a modulation-doped QW. For example, the band gap renormalization occurs due to the many-body effect, which leads to the band gap shrinkage. This phenomenon must be predominant at relatively low electron concentration.[5] Then, quenching of the excitonic absorption must be due to screening.[6,7] Apparently, it is this phenomenon that increases the absorption observed in the spectral region close to the band edge. One can also expect a certain Stark shift of the confined levels. Besides, the band mixing in the QW is supposed to be significant, because the optical transitions occur in a modulation-doped QW when the wave vector **k** is equal to the Fermi value rather than to zero, as in an undoped QW. This causes mixing of the light and heavy hole states and breaks the selection rules for the TE and TM waveguide polarizations. This suggestion can explain well the same spectral position of the lowest resonance and good separation of higher light and heavy hole states for the TE and TM waveguide modes. These states seem to provide smooth peaks in the absorption spectra due to the "forbidden" optical transitions $1hh \rightarrow 2e$ for the TE mode and $1lh \rightarrow 2e$ for the TM mode, which become allowed in the skewed QW. An insignificant voltage-induced shift of the peaks results mainly from the Stark shift of the confined levels.

A comparison has been made of the band edge shift measurements and the computation of the Moss–Burstein shift using the photocurrent data. The analysis confirms the domination of the blue absorption band edge shift due to the electron band filling when N_s is higher than $\sim 3 \times 10^{11}$ cm^{-2}.

Together with the excitonic absorption quenching, this phenomenon produces a significant electromodulation of absorption in the spectral range of about 60 meV around the band edge, and it is just this characteristic that gives rise to efficient active Q-switching in the low-threshold two-segment laser.

4. Two segment laser operation

Fig. 8 shows a schematic structure of a fabricated two-segment ridge laser. The ridge waveguide was chemically etched to a width of 12 μm. SiO$_2$ was used to confine the current injection to the ridge. The sections are electrically isolated and optically interconnected by a common waveguide. The electrical isolation of about 100 kΩ results from the 4 μm wide separation selectively etched between the segments.

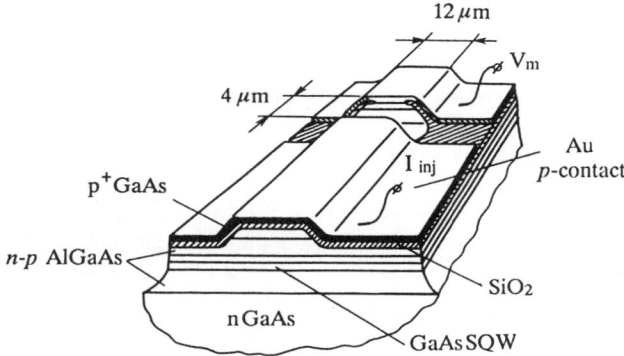

Fig. 8. Schematic view of the two-segment ridge laser.

The spectral width of efficient electrooptical modulation provided by free carrier effects is large enough to overlap even the electroluminescence region of the structure. That allows to test the modulation ability of the two segment structure in a small-signal electroluminescence operation, when the amplifier section is driven by current, much lower than the laser threshold current. Fig. 9(a) shows the intensity of the light-emission-diode radiation passed through the modulator and measured as a function of the reverse bias. This dependence again reflects the specific points 2 V and 4 V, which correspond respectively to the full overlap of the QW and the p-n junction SDR's and to the complete depletion of the QW. These values set up the limits for the voltage region of efficient electromodulation. They result only from the geometrical size and doping level and can be calculated before growing the structure. The structure is very suitable to study the electrooptical behaviour, while for the device application it is more convenient to shift the electromodulation region to the zero voltage. This idea has been realized in a similar structure. The modulation depth of the electroluminescence emission is about 75% for a 150 μm long modulator. So, the device is rather effective even in the light emission diode operation.

The threshold current density of the lasers with the modulator section of about 100 μm in length ranged from 400 to 800 A/cm^2 for the most transparent state of the modulator. The threshold current can be almost trebled by applying the DC bias to the modulator. In the active Q-switching operation the optical amplifier is driven by a continuous current, while a microwave signal is applied to the modulator. The DC bias can also be applied to the modulator if it is necessary to obtain efficient modulation. Only about 100 mV of the modulating voltage is necessary to provide stable Q-switching, so that every period of the microwave signal gives rise to a single optical laser pulse. The most important parameters of the laser generation are the pulsewidth and the attainable repetition rate. When the laser is long enough, the pulsewidth is limited by the entire laser cavity rather than by the modulator characteristics. The time dependence of the long laser generation is shown in Fig. 10. The total length of the laser cavity was 830 μm. The dependence was obtained using the p-i-n photodiode with time resolution of about 50 ps FWHM. The pulsewidth of about 160 ps and the upper limit of the repetition rate about 1.5 GHz seem to be determined by the dynamic characteristics of the entire laser cavity. For the 250 μm laser the pulsewidth is about 50 ps and seems to be limited by the photodetector time response. The repetition rate in this case is about 3 GHz and is very likely to be limited by the modulator frequency characteristics.

5. High-frequency characteristics of the modulator

Special measurements were made to find the frequency characteristics of the modulator in small-signal operation. The two-segment structures were used. The amplifier was pumped with relatively low current to prevent the laser generation in the system, which was, however, high enough to provide a probe light signal to be detected after passing through the modulator. A special technique was developed to achieve high time resolution of a small electroluminescence signal. For this, significant intrinsic nonlinearity of the modulator electrooptical response was used. The nonlinear character of the dependence of the output light intensity on the bias voltage (see Fig. 9(a)) allows to raise the sensitivity by conventional lock-in technique and a relatively slow photoelectron multiplier used as a light detector, while the frequency band of the set is determined by that of the modulator.

Applying a constant reverse voltage to the modulator we can change the output light signal according to the curve in Fig. 9(a). Superposition of the microwave sinus signal on the constant bias V_0 gives rise to a periodic output light signal. However, no change occurs in the registered signal, provided the V_0 is located within the linear portion of the modulator electrooptical response curve. This is due to application of the slow photomultiplier which integrates a high frequency signal. However, shifting the V_0 to the nonlinear portion of the curve increases or decreases the registered signal with respect to the level of the output signal when there is no microwave modulation (differential response), so that the latter reflects the nonlinearity of the modulator electrooptical response. Thus, by varying the V_0 and the microwave signal frequency one can measure the frequency response of the modulator.

In Fig. 9(b) the differential response measured at constant frequency is plotted against V_0. It can easily be seen that the curve corresponds to the electrooptical modulator response. Fig. 9(c) exhibits the frequency dependence of the response at constant V_0 corresponding to the extremum near 2.5 V. This dependence reflects the frequency bandwidth of the modulator and well accounts for the limitation on the repetition rate of the short laser Q-switching. The region of the efficient modulation extends approximately up to 3 GHz.

The high frequency characteristics of the device are determined by the rate of electron tunnelling through the QW barrier, provided the dimensions and, hence, the capacitance of the device are small enough. The rate is a function of the barrier height and width, which depend on the device design. The frequency characteristics can be improved through increasing the doping level,

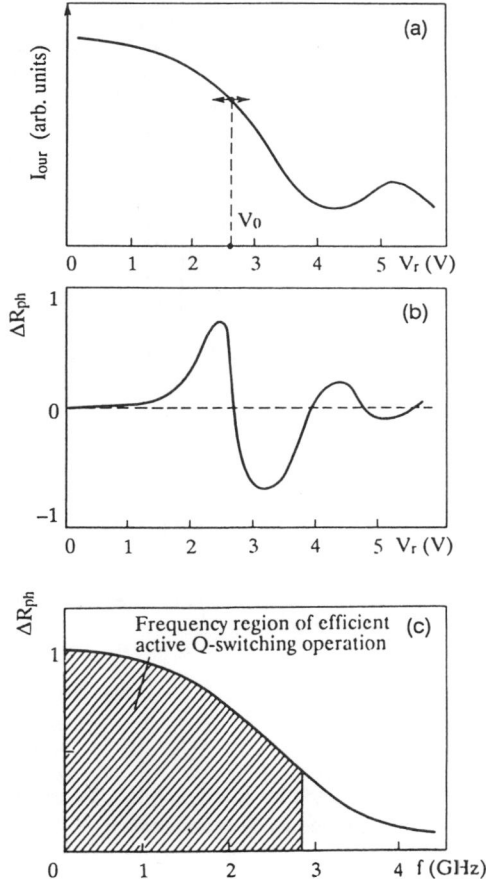

Fig. 9. Small-signal characteristics of the modulator. The modulator output (*a*) and differential response (*b*) against the reverse bias. Frequency dependence of the differential response at about 2.5 V of the DC reverse bias on the modulator (*c*).

which lessens the barrier width and height, increasing the tunnelling rate. However, the problem arises to meet two requirements: the modulator design providing high frequency characteristics and low threshold laser operation. The preliminary estimation shows the possibility of low threshold operation within the frequency band of about 5 GHz.

6. Conclusion

We have presented and discussed the measurements of electrooptical characteristics of a waveguide laser structure with a modulation-doped QW. It has been shown that the observed free carrier effects are very convenient to obtain efficient active Q-switching of a two-segment laser with very low threshold current density. The active Q-switching is achieved at the threshold current density as low as 400–800 A/cm² depending on the modulator length. Only 100 mV of the modulating microwave signal is needed to provide active Q-switching and to obtain optical pulses less than 50 ps wide with the repetition rate up to 3 GHz. It has been demonstrated that the frequency bandwidth of the short laser operation is determined by the high frequency characteristics of the modulator rather than by the dynamic features of the whole laser cavity.

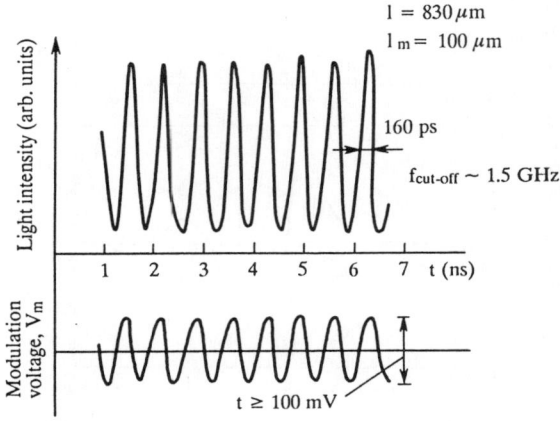

Fig. 10. Time response of the long laser operated by a 100 mV microwave signal.

Acknowledgments

The authors wish to thank Prof. R.A. Suris for many helpful discussions and Dr. V.S. Kalinovsky for expert technical assistance in high frequency measurements.

References

[1] D.Z. Tsang, J.N. Walpone, J.L. Liau, S.H. Groves, and V. Diadiuk 1984 *Appl. Phys. Lett.* **45** N 3 204–206
[2] Y. Arakawa, A. Larson, J. Paslaski, and A. Yariv 1986 *Appl. Phys. Lett.* **48** N 9 561–563
[3] S. Tarucha and H. Okamoto 1986 *Appl. Phys. Lett.* **48** N 1 1–3
[4] G. Livescu, D.A.B. Miller, D.S. Chemla, M. Ramaswang, T.Y. Chang, N. Sauer, A.C. Gossard, and J.H. English 1988 *IEEE J. Quantum electronics* **24** N 8 1677–1689
[5] I. Bar Joseph, J.M. Kuo, C. Klingshirn, G. Livescu, T.Y. Chang, and D.A.B. Miller, D.S. Chemla 1987 *Phys. Rev. Letters* **59** 1357
[6] H. Sakaki, H. Yoshimura, and T. Matsusue 1987 *Japan. J. Appl. Phys.* **26** 1104
[7] C. Tombling, M.M Stallard, and J.S. Roberts 1990 *Semicond. Sci. Technol.* **5** 502-507

Joint Soviet-American Workshop on the Physics of Semiconductor Lasers   May 20–June 3 1991

Degenerate sixwave mixing in the active region of a diode laser and a problem of lateral distribution stability

A.P. Bogatov

P.N. Lebedev Physical Institute, Academy of Science of the USSR,
53 Leninsky pr., 117924 Moscow, USSR

Abstract. A problem of the lateral field distribution stability in junction lasers with a wide active region is considered in terms of the degenerate sixwave mixing. It is shown that a periodic modulation of the field distribution arises due to the optical nonlinearity. Threshold of the lateral instability depends essentially on the coefficient of amplitude-phase coupling.

Blue semiconductor laser research at the University of Florida

P.S. Zory
University of Florida, Gainesville, FL, USA

Abstract. In October 1988, a research program was initiated at the University of Florida (UF) with the goal of developing epitaxial diode structures capable of efficient light emission in the blue-green region of the electromagnetic spectrum. Devices such as semiconductor lasers fabricated from such material would be of considerable value in areas such as high density optical storage and high definition color displays. Although diode lasers have not yet been demonstrated, considerable progress has been made in showing that ZnSe is a very good candidate for room temperature diode laser action at 470 nm. For example, room temperature photo pumped lasing was demonstrated for the first time in epitaxial thin films of ZnSe grown by MBE and MOCVD on GaAs substrates. Although GaAs is very absorbing at 470 nm, the actual waveguide losses were small leading to the possibility of developing efficient, antiguide diode light emitters. Also demonstrated were ZnSe:N/ZnSe:Cl *p-n* homojunction light emitting diodes fabricated using a novel nitrogen atom beam doping procedure during MBE growth. These and other results achieved in the UF blue diode laser program will be reviewed.

Tunable diode lasers for 3 to 40 μm infrared spectral region

A.P. Shotov

P.N. Lebedev Physical Institute, Academy of Sciences of the USSR,
53 Leninsky pr., 117924 Moscow, USSR

Abstract. Operating parameters and structures of typical diode lasers based on the ternary lead-salt crystal systems PbSnSe, PbSSe and PbCdSe are presented. We discuss the characteristics of traditional diffusion laser structures and some improved structures with controlled profiles of carrier concentration and SC SQW heterostructures prepared by hot wall molecular epitaxy.

1. Introduction

Diode lasers fabricated from IV–VI narrow gap lead-salt semiconductor compounds and alloys are used as current- or temperature-tunable sources in the 3–40 μm spectral region for high resolution spectroscopy, monitoring of atmospheric pollution, gas analysis and other applications.[1,2] The IV–VI lead-salt family of semiconductors includes PbS, PbSe(Te) and their alloys with SnSe, CdS, EuS, EuSe, and SrSe. The bandgap E_g of lead salts occurs at the L-point of the Brillouin zone (instead of the Γ-point for III–V compounds) and varies from 0.3 eV to zero, depending on the composition of the alloy. Lasers made of these alloys cover 3 to 40 μm infrared spectral region. These types of laser usually require cryogenic cooling from Dewar or modern closed cycle He gas refrigerators, which allows to vary the laser temperature from about 10 K to room temperature. Thermoelectric cooling can be used for diode lasers of 3–8 μm spectral region pulse operated at higher temperature ($T > 200$ K).

2. General characteristics of lead-salt tunable diode lasers

The main problem for lead-salt diode lasers is an improvement of the operating temperature, the threshold current and the long time stability. During last few years much progress has been made in the design and performance of the lasers. Different types of heterostructures, efficient optical and carriers confinement, extremely thin quantum well active layers have been used together with new compositions of the alloys (PbSrSe, PbEuSe). Now the highest operating temperature up to 250 K can be reached for the IR spectral region short wavelengths (3–6 μm). A typical value of the threshold current density is $J_{th} = 0.5 - 2$ kA/cm^2 at 80 K. Emission power of the order of $0.1 \div 10$ mW is usual. This power together with a narrow linewidth (10^{-5} cm^{-1}) gives very high spectral brightness, which is a very important characteristic for many spectroscopic applications. For diode lasers with Fabry–Perrot optical cavity length L, the separations between the longitudinal emission laser modes are:

$$\Delta\left(\frac{1}{\lambda}\right) = \frac{1}{2L(n-\lambda)} \qquad (1)$$

where the effective refractive index is

$$n^* = n - \lambda \frac{dn}{d\lambda}$$

For most alloys $n^* \sim 5 \div 6$, and for a typical laser cavity length $L \sim 0.4$ mm $\Delta(1/\lambda) \sim 2$ cm^{-1}. The spectral region of gain in these materials is rather wide, and several laser modes are usually observed simultaneously. Emission wavelengths of the lasers can be tuned during the operation by varying some of the parameters being dependent on the bandgap and the refractive index.[3-5] The most widely used is both the temperature and current control tuning. The frequency range where

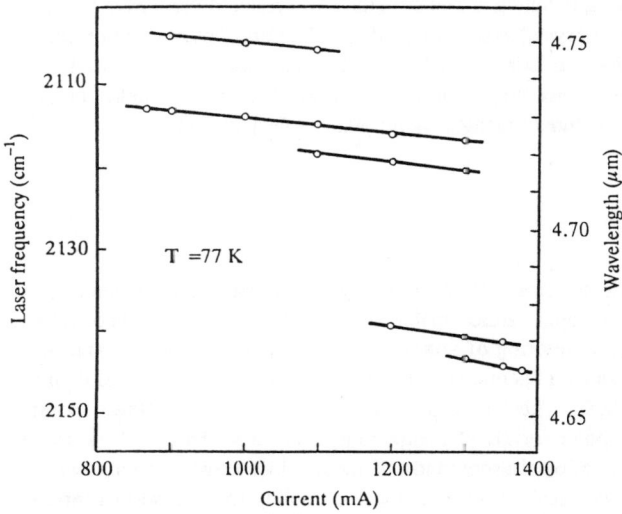

Fig. 1. Spectral mode frequencies for PbS$_{0.65}$Se$_{0.35}$ diode laser versus diode current.

an individual spectral mode can be continuously tuned with the use of current or temperature tuning (fine tuning) is usually limited by 1–2 cm^{-1} (Fig. 1). A single mode tuning range as high as 15 cm^{-1} was observed in some of PbSSe diode lasers.[6] The rate of change of emission frequency with laser current depends on the variation of both the refractive index and the optical cavity length with temperature

$$\frac{d\lambda}{dT} \cdot \frac{dT}{dI} = \frac{\lambda}{n^*} \left(\frac{dn}{dT} + \frac{n}{L} \frac{dL}{dT} \right) \frac{dT}{dI} \qquad (2)$$

Heating of the diode crystal by current is determined by thermal resistance of the devices and by electrical contact resistance. Current and temperature tuning rates < 0.5 cm^{-1}/mA and < 1 cm^{-1}/K are usually observed. Quasi-continuous wide tuning of mode frequencies is determined by the shift of the optical gain region with temperature due to dE_g/dT and dn/dT (Fig. 2). The maximum tuning range depends on the operating temperature range ΔT of the diode laser. A typical value of the tuning range is $\delta(1/\lambda) \simeq 100 - 300$ cm^{-1}. A very important problem is thermal stability of the laser devices. For most lead-salt materials used in diode lasers $dE_g/dT > 0$ (instead of $dE_g/dT < 0$ for III–V materials). This means that all nonuniformities of the p-n junction and current density are damped which improves the thermal stability of the diode lasers. The reliability of the lasers is mainly associated with thermal cycling degradation of the laser operating characteristics. In our work some stable laser structures have been developed and good and stable multilayer (Au-Pd-In) electrical contacts have been made. The next Section summarizes the most representative types of diode lasers.

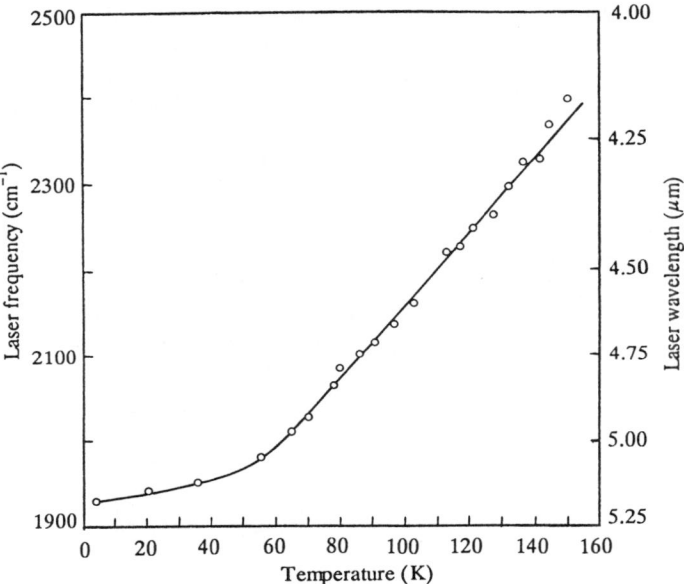

Fig. 2. Temperature dependence of the center frequency of emission for $PbS_{0.65}Se_{0.35}$ laser.

3. Diffusion lasers

The ternary compounds PbCdS, PbSSe and PbSnSe are used for diffusion-based technology to cover the 3–40 μm spectral range. The crystals were grown using a closed-tube directional vapour-phase technique, described previously.[7] To prepare the *p-n* junction, *n*-type crystals with carrier concentration from $5 \cdot 10^{17}$ to $5 \cdot 10^{18}$ cm^{-3} were used. The type of conductivity and carrier concentration of IV–VI semiconductors can be controlled by small deviation from the stoichiometry of the material. The *p-n* junction was formed by interdiffusion process from the vapour phase using a Se-rich $Pb_{0.49}Se_{0.51}$ source. Substrate *n*-type wafers and a $Pb_{0.49}Se_{0.51}$ source are sealed in an evacuated quartz tube. Heating of the tube for 0.5–1 h at 400–450 °C (depending on the material) creates a *p-n* junction at a depth of 10–15 μm.[1,3–6] A multilayer (Au-Pd-In) plating was used to make electrical contacts to the *p*- and *n*-sides of the *p-n* junction. Stripes of 150 μm wide were separated and laser end faces were formed by cleaving. The distance between the end faces usually varied from 400 to 500 μm. The devices were packaged by In-welding to a In-plated heat sink. Typical threshold current density of the lasers was of the order of 0.1 kA/cm^2 at 4.2 K, and $1 \div 3$ kA/cm^2 at 80 K. This type of diffusion lasers has demonstrated good thermal stability and long time reliability and is now used in most spectroscopic applications.

4. PbSSe diode lasers with controlled carrier concentration

Higher operating temperature has been achieved by using a double heterostructure.[8,9] However the lattice mismatch of the DH becomes often the cause of degradation after subsequent thermal cycling between low and room temperature. The problem of lattice mismatch can be avoided, and efficient confinement can be simultaneously achieved in a homostructure with controlled carrier concentration profiles.[10] The carrier and optical confinement is enhanced due to the potential barrier of the n^+-p-p^+ structure and to a fairly strong carrier concentration dependence of the refractive index of the narrow gap IV–VI semiconductors.[11,12] PbSSe diode lasers with controlled carrier concentration profile cover the infrared spectral region from 4 to 8 μm, the most useful for monitoring a great variety of air pollutants, gas analysis and other applications. The possibility of optical confinement from the carrier concentration profile can be seen from Fig. 3. The Figure shows refractive index versus carrier density for $PbS_{0.65}Se_{0.35}$ at photon energy close to the energy gap. For the concentration step from 10^{17} to $2 \cdot 10^{18}$ cm^{-3} a corresponding refractive index step $\delta n/n \sim 10\%$ can be obtained, which is large enough for optical confinement. Calculations of the confinement vector Γ show that 80% of the radiation should be located in a 1.5–2 μm active layer.

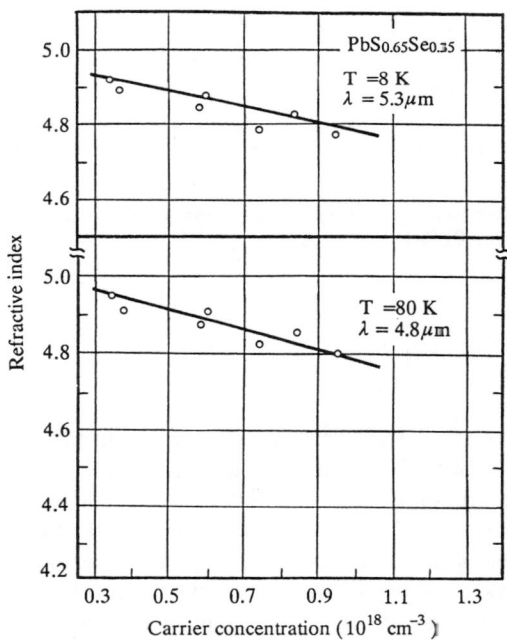

Fig. 3. Dependence of the refractive index in $PbS_{0.65}Se_{0.35}$ on carrier concentration.

Calculations of the relative positions of the Fermi and quasi-Fermi levels (Fig. 4) show that the potential barriers of the n^+-p-p^+ structure (with the carrier concentration $1 \cdot 10^{17}$ cm^{-3} in the active layer and $2 \cdot 10^{18}$ cm^{-3} in the n^+ and p^+ layers) can be ≈ 10 kT at 4.2 K and several kT at 80 K. This is quite sufficient for carrier confinement.

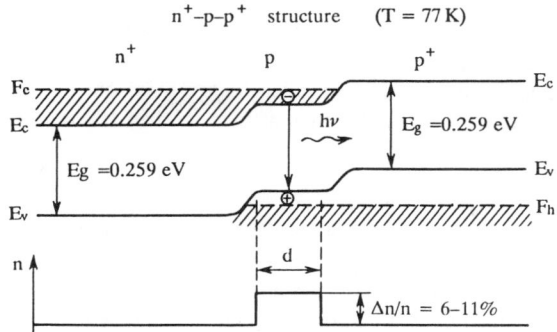

Fig. 4. Scheme of n^+-p-p^+ laser structure at forward bias demonstrating carrier and optical confinement.

The laser n^+-p-p^+ homostructure (Fig. 5) was grown by hot wall molecular epitaxy.[13] PbSSe n^+ substrate about 100 μm thick and with carrier concentration $2 \cdot 10^{18}$ cm^{-3} was grown on KCl wafer oriented in the [100] direction. KCl was dissolved by water after the growing process. All the subsequent layers were grown on KCl side of PbSSe substrate. A good quality n^+-layer ($n = 2 \cdot 10^{18}$ cm^{-3}) about 10 μm thick, a n-p junction and an active p-layer were grown at a rate of 3 μm/h by proper adjustment of the compensating selenium vapour pressure. A contact Tl-doped p^+-layer with the carrier concentration of about $2 \cdot 10^{18}$ cm^{-3} and thickness 5–10 μm was grown during the second deposition step. The system remained under vacuum throughout the whole process of growth, also between the two deposition steps.

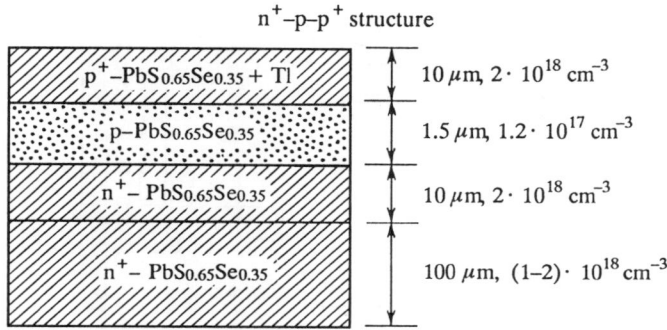

Fig. 5. Cross section of a n^+-p-p^+ laser structure grown by hot wall molecular epitaxy.

The best results were obtained for an active 1.5 μm p-layer with the carrier concentration $1.2 \cdot 10^{17}$ cm^{-3}, estimated from the C–V characteristics of the p-n junction. Low carrier concentration of the active layer is very important to decrease the free carrier absorption. The threshold current density for the best PbS$_{0.65}$Se$_{0.35}$ lasers was about 25 A/cm^2 at 4.2 K and 700 A/cm^2 at 80 K.

Fig. 2 shows the center emission frequency of the Pb$_{0.65}$Se$_{0.35}$ diode laser as a function of temperature. The temperature tuning range is comparatively large: about 500 cm^{-1} ($\lambda = 5.3 \div 4.1$ μm) for 4.2–150 K range. Fig. 1 illustrates how individual spectral modes vary with change of the diode current. Each mode has a tuning range 1–3 cm^{-1}. A current tuning rate of about 10^{-2} cm^{-1}/mA was usually observed. The maximum power of 22 mW at 4.2 K was obtained. This value corresponds to the internal differential quantum efficiency of about 20% which is rather high for these laser types.

The temperature stability of the laser is much higher than that of DH lasers. Most of the lasers showed no changes in the threshold current after more than 100 temperature cycles between 300 and 4.2 K.

5. PbSSe/PbSnSe heterostructure lasers with a quantum-well active region

To improve the threshold current and operating temperature PbS/PbSSe/PbSnSe double heterostructures with separate electron and photon confinement and with a quantum well in the active region (SC SQW DH's) were fabricated by hot wall molecular epitaxy.[13-16] The composition of seven layers of the structure is shown in Fig. 3. The single-quantum $Pb_{0.95}Sn_{0.05}Se$ active region and the two $PbS_{0.4}Se_{0.6}$ wave-guiding layers near the active region result in an increase in the energy bandgap of $\Delta E_g = 109$ meV at 77 K. The n-type layers were doped with bismuth. Selenium concentration was adjusted to obtain p-type conductivity. The quantum-well active region was undoped. The thickness (L_z) of the active region was 400, 500, 1000, 2000 Å and up to 4 μm.

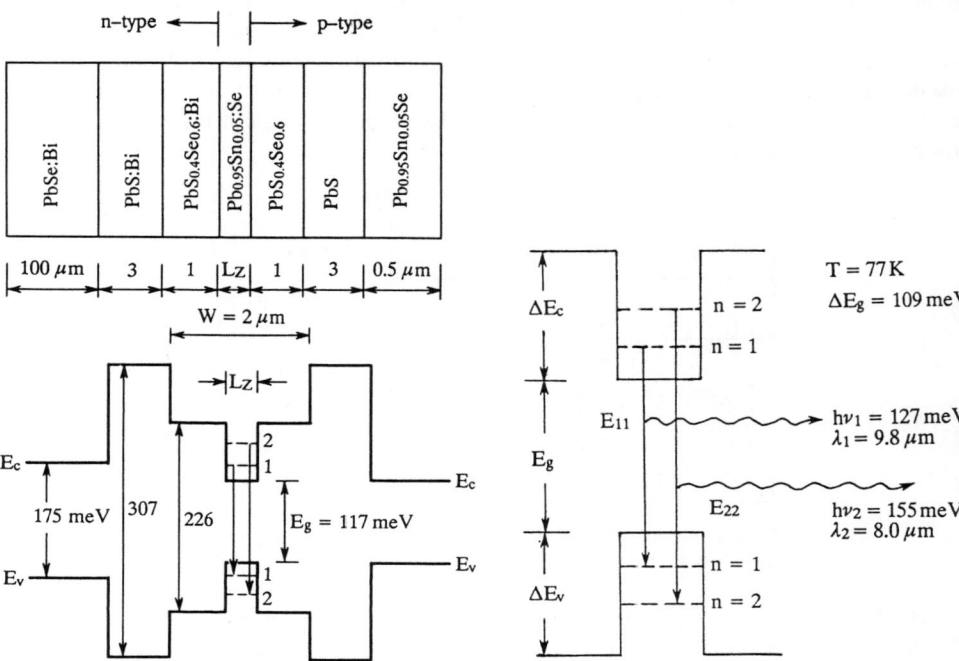

Figure 6. Schematic energy band diagram and laser structure with single-quantum-well PbSnSe active region.

Figure 7. Experimentally measured emission energies.

We have found that the photon energy for the lasers with $L_z > 1000$ Å corresponds to the bandgap E_g of the active region. For the lasers with $L_z = 400$ and 500 Å two different laser emission lines with energies greater than E_g were observed. For $L_z = 400$ Å this increase is as high as 9.7 meV for the line with $h\nu_1 = 127$ meV ($\lambda = 9.8$ μm) and 38 meV for the line with $h\nu = 155$ meV ($\lambda = 8$ μm) at 77 K (Fig. 7). These shifts are found from the difference between the photon energy of the QW laser and that of the DH laser with a thick active region ($L_Z \geq 1$ μm) in pulsed operation (to eliminate heating effects). The energy levels in the quantum well ($L_Z = 400$ Å) were calculated from the model square potential well of finite depth. Since the exact value of the band discontinuities in PbSnSe/PbSSe is unknown, we use $\Delta E_c = \Delta E_\nu$. In our case, $\Delta E_g = 109$ meV and $m_e = 0.35 m_0$, $m_h = 0.031 m_0$. The observed photon energies were estimated to correspond to optical transitions between the localized states in the potential well with $n = 1$ (E_{11}) and $n = 2$ (E_{22}). The selection rule $\Delta n = 0$ holds for these transitions. (The fact that the effective masses of the electrons and holes are very nearly equal leads to a weak dependence of E_{11} and E_{22} on $\Delta E_c/\Delta E_\nu$). Similar results were observed for the laser with $L_Z = 500$ Å.

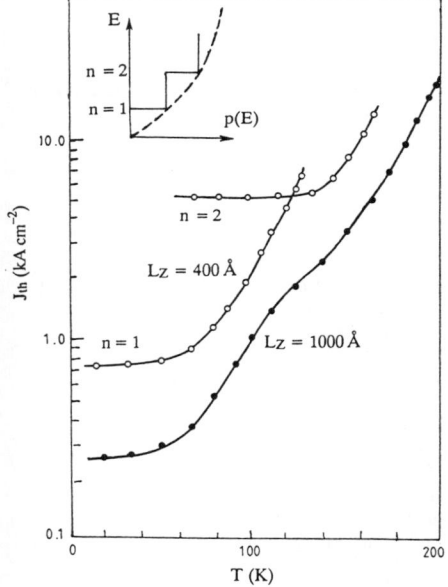

Figure 8. Temperature dependence of the threshold current density.

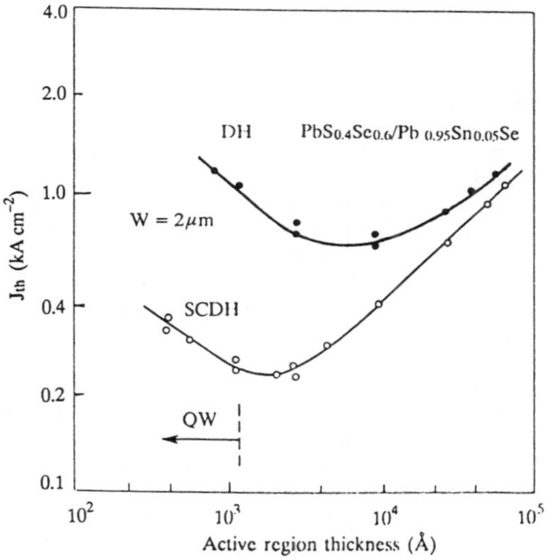

Figure 9. Threshold current density as a function of active layer thickness at 77 K.

Fig. 8 shows the temperature dependence of the threshold current density J_{th} for the lasers with $L_Z = 400$ and 1000 Å with pulsed pumping (1 μs, 600 Hz). At low temperature, $T < 60$ K, the laser with $L_Z = 400$ Å operates at low J_{th} for optical transitions between the $n = 1$ states ($h\nu_1 = E_{11}$), because only the states which are close to the band edge are filled by carriers. Transitions between the $n = 2$ states require higher pumping current. As was noted above, at $T = 60 - 120$ K two laser emission lines were observed simultaneously. At $T = 125$ K only the $h\nu_2 = E_{22}$ transition was observed because of higher density of the $n = 2$ states. In part of the temperature range, the threshold current for the transitions E_{11} and E_{22} was practically independent of the temperature. This fact reflects step-wise shape of the density of states for the quasi-2D system in the potential well. Similar behavior is also found for the laser with $L_Z = 500$ Å. For the laser with $L_Z = 1000$ Å the $J_{th}(T)$ curve has a structural feature which indicates a switching of the laser operation from the $n = 1$ to $n = 2$ optical transitions. However, the photon energy in this case is quite close to E_g of the active layer. The temperature dependence of the threshold current for the laser with $L_Z = 2000$ Å is of standard type for lead-salt diode lasers.[14]

Fig. 9 shows the relation between the threshold current density and the active layer thickness for these SC DH lasers and also for the DH $PbS_{0.4}SE_{0.06}/PbS_{0.05}Se$ at $T = 77$ K. The effect of separate confinement for carriers and photons is clearly demonstrated. The lowest threshold current of 230 A·cm^{-2} at 77 K and the highest operating temperature of 218 K (for pulsed regime, $\lambda = 6.5$ μm) were obtained for the SC DH laser with $L_Z = 2000$ Å. In the single-quantum-well lasers with $L_Z < 1000$ Å the threshold current increases with decreasing thickness of the QW active region. We suppose that the main reason for it is a carrier leakage from the potential well due to small values of the ΔE_g in these structures. A larger gap material (like PbSrSe) for barrier layers can be useful in this case.

Acknowledgments

The results presented here have been obtained in collaboration with my colleagues at the Laboratory of Narrow-gap Semiconductors of the Lebedev Physical Institute.

References

[1] A.P. Shotov 1973 *Jpn. J. Appl. Phys.* **42** 282
[2] A.P. Shotov 1986 *Vestnik Ak. Nauk SSSR* **6** 3
[3] I.I. Zasavitsky, B.N. Matsonashvili, V.I. Pogodin, and A.P. Shotov 1974 *Fiz. Tekh. Poluprovodn.* **8** 732
[4] I.I. Zasavitsky, B.N. Matsonashvili, and A.P. Shotov 1975 *Pis'ma Zh. Tekh. Fiz.* **1** (7) 341
[5] I.I. Zasavitsky, E.G. Chizhevsky, and A.P. Shotov 1978 *Kvant. Elektron.* **5** (3) 692
[6] M.S. Murashov and A.P. Shotov 1990 *Kvant. Elektron.* **17** (12) 2426
[7] A.P. Shotov, I.V. Kucherenko, Yu.N. Korolev, and E.G. Chizhevsky 1972 *Fiz. Tekh. Poluprovodn.* **6** (8) 1508
[8] A.P. Shotov, K.V. Vyatkin, and A.A. Sinyatinskii 1980 *Pis'ma Zh. Tekh. Fiz.* **6** 983
[9] A.P. Shotov, K.V. Vyatkin 1980 *Pis'ma Zh. Tekh. Fiz.* **6** 1199
[10] A.P. Shotov and A.A. Sinyatinskii 1983 *Pis'ma Zh. Tekh. Fiz.* **9** 881
[11] K.V. Vyatkin and A.P. Shotov 1980 *Fiz. Tekhn. Poluprovodn.* **14** 1331
[12] A.A. Sinyatinskii and A.P. Shotov 1980 *Fiz. Tekh. Poluprovodn.* **16** 2187
[13] K.V. Vyatkin, A.P. Shotov, V.V. Ursaki 1981 *Izv. Akad. Nauk SSSR, Neorg. Mater.* **17** 24
[14] A.P. Shotov and Yu.G. Selivanov 1986 *Pis'ma Zh. Tekh. Fiz.* **12** 1386
[15] A.P. Shotov and Yu.G. Selivanov 1987 *Pis'ma Zh. Eksp. Teor. Fiz.* **45** 5
[16] A.P. Shotov and Yu.G. Selivanov 1990 *Semicond. Sci. Technol.* **5** 927

InGaSbAs/GaAlSbAs heterostructures for mid-infrared injection lasers

B.N. Sverdlov

P.N. Lebedev Physical Institute, Academy of Science of the USSR,
53 Leninsky pr., 117924 Moscow, USSR

Abstract. It was shown earlier that quaternary InGaSbAs/GaAlSbAs heterostructures lattice-matched to GaSb could be used for fabrication injection lasers and LED's, emitting at the wavelength beyond 1.8 μm. Here we discuss here the optical structure and band diagrams of these heterostructures. The problems of liquid phase epitaxy of InGaSbAs and GaAlSbAs quaternaries including the problems of epitaxial growth on profiled substrates are also considered. The main characteristics of the injection heterolasers operating over the wavelength range of 1.8–2.4 μm are presented. The properties of lasers with a broad contact and those of stripe lasers at continuous wave operation are discussed.

Extremely smooth AlGaAs-GaAs quantum wells grown by metalorganic chemical vapor deposition

R.D. Dupuis, J.G. Neff, and C.J. Pinzone

Microelectronics Research Center and NSF Science and Technology Research Center, The University of Texas at Austin
Austin, Texas 78712-1084, USA

Abstract. We report here the low-pressure MOCVD growth of AlGaAs-GaAs quantum well heterostructures having low-temperature (4.2 K) photoluminescence spectra with full-width-at-half-maximum (FWHM) values ranging from ~ 6 − 4 meV for quantum wells having 6–28 monolayer (ml) widths, respectively. These linewidths are compared to those measured for quantum wells grown by molecular beam epitaxy, flow-rate modulation epitaxy, and atomic layer epitaxy. We find that the FWHM values for the thinnest wells grown in the present study (~6 ml) are equal to or narrower than those observed for comparable structures by other technologies.

1. Introduction

AlGaAs-GaAs quantum-well (QW) heterostructures were first grown[1] by metalorganic chemical vapor deposition (MOCVD) in 1978. These structures were of very high quality and yielded the first room-temperature QW injection lasers.[1] While the optical quality of MOCVD QW heterostructures is known to be generally excellent, the interface smoothness and abruptness of MOCVD QW's has not been shown to be equal to the best reported values for similar structures grown by MBE. In particular, the full-width-at-half-maximum (FWHM) of the low-temperature photoluminescence (PL) spectra of MOCVD QW's is typically > 2 times that of similar structures grown by optimized MBE growth techniques. It has been shown that interrupting the growth during the MBE deposition of AlGaAs-GaAs QW's can greatly improve the smoothness of the QW interfaces.[2] Continuous MBE growth is known to produce AlGaAs-GaAs interfaces that are somewhat rough.[2] Interruptions of 90–120 s at each interface during MBE growth can produce QW's with large regions of monolayer (ml) smoothness.[2,3] MBE-grown QW's using low V/III ratio and low growth temperatures and interrupted growth exhibited very narrow PL lines at temperatures of ~ 2 − 6 K and distinct peaks associated with wells exactly one monolayer different were observed. The narrowest FWHM observed for ~6 ml well was ~6 meV.[2] Linewidths in this range have been associated with QW's having smooth interfaces over the ~250 Å extent of the exciton diameter.[2,3] MOCVD-grown AlGaAs-GaAs QW's in this thickness range generally exhibit FWHM values >12 meV.[4,5]

We report here the MOCVD growth of AlGaAs-GaAs multiple-quantum well (MQW) heterostructures having low-temperature (4.2 K) PL spectra with FWHM values ranging from ~6–4.0 meV for QW's having 6–28 ml widths, respectively. These values are less than that previously reported for conventional MOCVD-grown AlGaAs-GaAs QW's and also less than the linewidths previously reported for similar structures grown by interrupted growth schemes such as flow-rate-modulation epitaxy (FME)[6] and atomic-layer epitaxy (ALE)[7] using metalorganic precursors.

2. Crystal growth

The QW heterostructures reported here are grown at low-pressure (60 Torr) on GaAs (100) substrate using an Emcore GS3200 growth system. This system employs a 5 in. diameter Mo wafer carrier that is rotated at high speeds and has a capacity of three 2-in. diameter wafers. The stainless-steel growth chamber is accessed through a stainless-steel loadlock that contains a multiple-wafer-carrier cassette to allow three sequential runs to be made without opening the load lock to atmosphere. The alkyl source modules employ individual pressure balancing and temperature control and are coupled to a pressure-balanced fast-switching injection manifold. The sources employed were trimethylgallium (TMGa), trimethylaluminum (TMAl), and 100% arsine (AsH_3). The TMGa was held at a bath temperature of $-15\,°C$ and the pressure was controlled to 600 Torr. The TMAl was used at the same pressure but the source temperature is $20\,°C$. Purified H_2 was used as a carrier gas. Growth temperatures were varied in the range $700 - 800\,°C$ and growth rates in the range $\sim 0.5-1$ ml/s were used. The alloy composition, x, of the $Al_xGa_{1-x}As$ barrier was determined from the PL spectra and was varied in the range $0.23 < x < 0.30$. Typically, four GaAs QW's were grown having nominal thickness of 6, 12, 18 and 24 ml corresponding to growth times of 0.1, 0.2, 0.3, and 0.4 min. The multiple-quantum-well (MQW) structures have ~ 300 Å AlGaAs barriers with the alloy composition, x, in the range $0.23 < x < 0.30$. The V/III ratio (for the GaAs QW's) was varied between 25–120. The mole fraction of TMGa and TMAl was typically $\sim 3 \times 10^{-5}$ and the AsH_3 mole fraction was $\sim 4 \times 10^{-3}$. The rotation rates used in this study varied from 800–1400 rpm. The GaAs substrates were oriented to within $0.1°$ of the (100) plane and were degreased in solvents and briefly etched in 10:1:1 $H_2SO_4 : H_2O_2 : H_2O$ and rinsed in deionized water prior to growth.

3. Photoluminescence spectra

The PL spectra were taken in a variable-temperature Janis Super Vari-Temp liquid-He cryostat using the 5145 Å line of an Ar ion laser operating CW with an output power of 10 mW. The laser output was not focused and the excitation level at the sample surface is estimated to be $\sim 100\,mW/cm^2$. While the spectra were taken, the MQW samples were immersed in the liquid He and the temperature was held to about 4.2 K. Shown in Fig. 1 is the PL spectrum of a MQW wafer having AlGaAs ($x = 0.25$) barriers and grown at $730\,°C$ using a growth rate of ~ 1 ml/s and a V/III ratio of ~ 120. As shown in Fig. 1, the FWHM values for the luminescence from the QW having a thickness of 6 ml, 13 ml, 21 ml, and 28 ml are 7.7, 6.5, 6.1, and 6.1 meV. The data of Fig. 2 are for a MQW sample with $x = 0.22$ AlGaAs barriers but grown at $750\,°C$. These QW's are of thickness 6, 14, 21, and 28 ml and have FWHM values of 6.3, 5.3, 5.8, and 4.8 meV. In Fig. 3, PL data are shown for a similar structure grown at $750\,°C$ but having AlGaAs ($x = 0.27$) barriers. The FWHM values are 8.4, 6.5, 4.8, and 4.0 meV for QW's of 7, 14, 22, and 31 ml, respectively. These data are representative of the MQW wafers grown under these conditions. It has been found that the FWHM is lowest for QW's grown in the temperature range $750 - 780\,°C$, as shown by the data of Figs. 4 and 5. Fig. 4 shows MQW PL spectra for three wafers grown at 700, 730, and $750\,°C$. Fig. 5 shows similar data for wafers grown at 750, 780, and $800\,°C$. It is expected that the alloy composition of the barriers will affect the FWHM of the PL observed from the QW.[8] Although the alloy composition of the barriers varies in these wafers, the trend to narrower FWHM's for the higher growth temperature is demonstrated by these data. From the data of Fig. 5, we conclude that extremely sharp QW luminescence can be obtained from MQW structures grown at temperatures in the range 750 – 800 °C which are required for the MOCVD growth of high-quality AlGaAs. This is, of course, desirable for the growth of high-performance optoelectronic device structures.

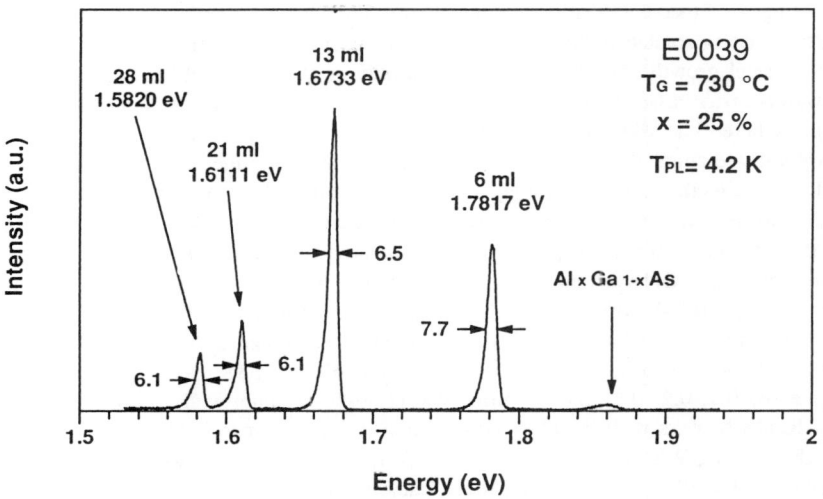

Fig. 1. Low-temperature (4.2 K) PL spectrum of 4-well MQW heterostructure grown at 730 °C by MOCVD. The AlGaAs barrier composition is $x = 0.25$.

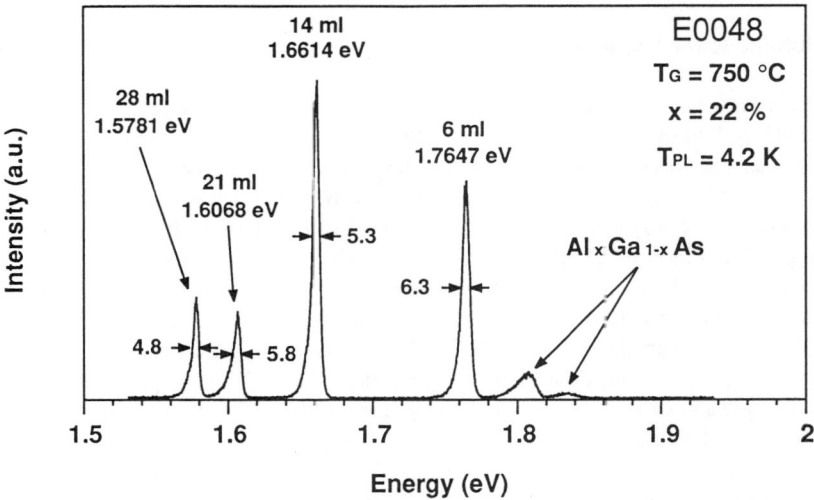

Fig. 2. Low-temperature (4.2 K) PL spectrum of a 4-well MQW heterostructure grown at 750 °C by MOCVD. The AlGaAs barrier composition is $x = 0.22$.

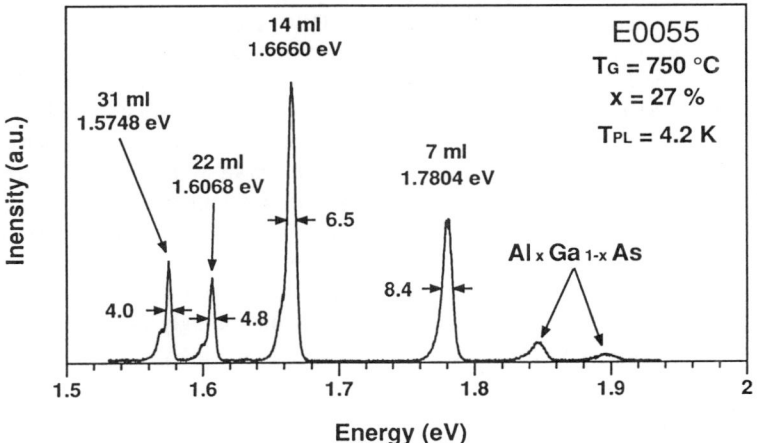

Fig. 3. Low-temperature (4.2 K) PL spectrum of a 4-well MQW heterostructure grown at 750 °C by MOCVD. The AlGaAs barrier composition is $x = 0.27$.

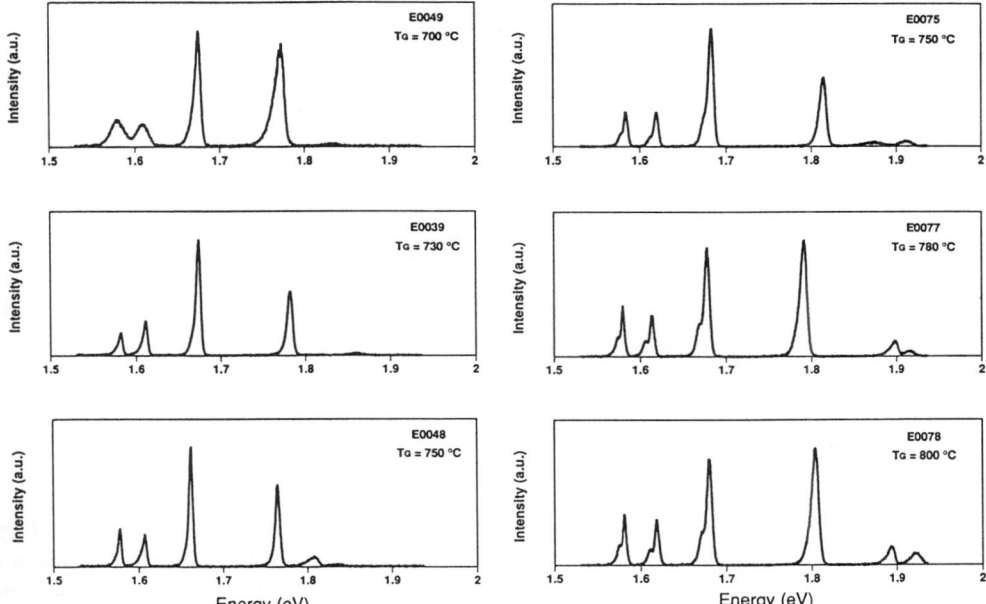

Figure 4. PL spectra of MQW wafers grown at 700, 730, and 750 °C.

Figure 5. PL spectra of MQW wafers grown at 750, 780, and 800 °C.

This is in marked contrast with the results of Miller et al.,[9] for MBE-grown MQW heterostructures. In their work, continuous growth at temperatures of 600 °C produced thick QW's (~20 ml) having nearly the optimum FWHM values (comparable to the FWHM values obtained for interrupted growth) while similar ~20 ml wells grown at 680 °C had very broad and weak PL emission.[9] It is widely observed that the highest quality MBE AlGaAs is grown at temperatures above 650 °C.[10] As a result, it appears that smoothest QW's are produced by MBE at conditions that are very unfavorable for the growth of high-quality AlGaAs.

It has been previously reported that the PL FWHM values of MQW wafers grown by MBE are very depended upon the growth interrupt time.[11] Growth times of 30 s do not produce very sharp PL from wells even for wells in the ~25 ml range.[11] Tu, et al.,[2,9] have used 120 s interrupts at each interface to produce very sharp QW PL emission for MBE-grown MQW wafers. These relatively long interrupt times are not desirable in that it is possible that impurities (e.g., O or C) have a long exposure times to the "exposed" QW interfaces, possibly leading to reduced carrier mobility and degraded quantum efficiency of QW heterostructure devices. In the present study, the interrupt time was varied between 0 and 120 s. While the best results to date have been obtained for 6 s interruptions, the linewidth for the wafer of this study is not strongly affected by the interrupt times in the range 6–120 s between layers.

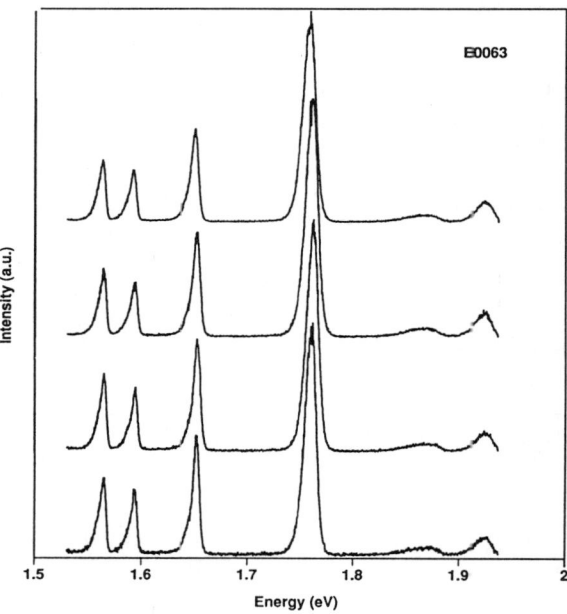

Fig. 6. Low-temperature PL spectra at four different positions on a wafer. The linear dimension scanned is ~8 mm.

It has also recently been observed for MBE-grown wafers that the position of the sharp (FWHM ~1 meV) QW emission peaks for thicker QW's can vary over a sample in a manner that is not consistent with the interpretation that the QW interfaces are atomically smooth and uniform on a monolayer scale over the dimension of an exciton.[12] These data were taken at 2 K with an excitation diameter of 0.2 mm. Moving this beam at 1 mm intervals gave PL data that showed anomalous variations in the QW dimension.[12] This implies that the PL spectra may be dependent upon the position of the PL excitation beam on the wafer and also the size (diameter) of the excited region. In our study, we have used a large excitation beam (~2 mm diameter) to take the PL data. Any nonuniformities in quantum well characteristics over these dimensions would contribute to the widening of the PL spectra. We do not observe unusually large FWHM values for the narrowest wells, so we conclude that the effects observed by Warwick, et al., for MBE grown QW's[12] do not exist in our wafers. In addition, in one case, we loaded a large sample (~10 mm in length) in the He cryostat (limited by the size of the copper sample mount) and scanned the laser excitation spot along the wafer to measure the uniformity of the PL emission. Tha data are shown in Fig. 6 for a MQW wafer grown at 750 °C. The individual spectra were taken about 2 mm apart so this figure represents data over a linear sample dimension of ~8 mm. The spectra are essentially identical and we conclude that the QW's are very uniform in thickness across this portion of the wafer.

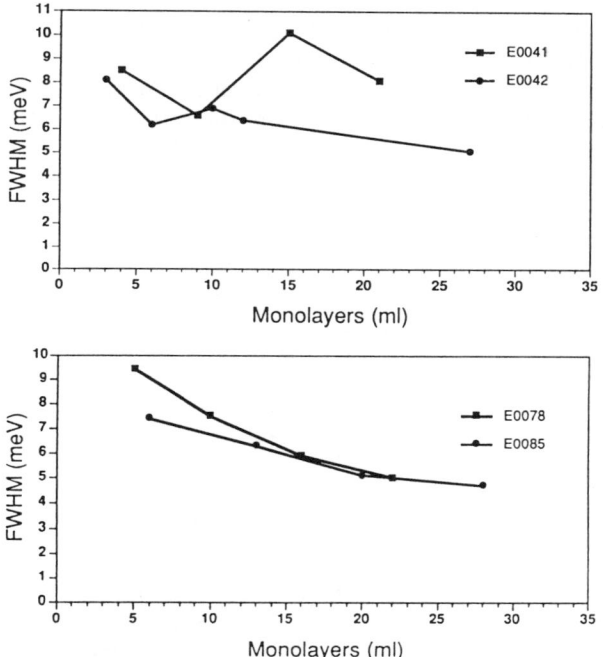

Fig. 7. FWHM of QW luminescence vs. thickness for wafers grown at ~1 ml/s (squares) and ~0.5 ml/s (circles).

The MBE-grown quantum wells that have shown narrow PL lines have been grown at growth rates in the range 0.25–0.5 ml/s.[2,13] In conventional MOCVD, the GaAs growth rates normally used are ~5–10 ml/s. In order to compare the results obtained from the MOCVD-grown wafers of our study with the published data on MBE-grown MQW wafers, we lowered the growth rate to 0.5 ml/s and 1 ml/s for this study by reducing the TMGa and TMAl partial pressures during growth. Shown in Fig. 7 (upper box) are data taken on wafers grown at ~1 ml/s (E0041, squares) and at ~0.5 ml/s (E0042, circles) with a constant V/III ratio of ~120. Both samples had AlGaAs barrier layers with x ~0.22 and were grown at 750 °C using 6 s interrupts at each interface. The QW's grown at the lower growth rate of ~0.5 ml/s show a reduced FWHM for all QW thickness. In the lower box of Fig. 7 are data for two other MQW wafers grown at 1 ml/s (E0078, squares) and 0.5 ml/s (E0085, circles) grown at 750 °C using 6 s interrupts but with AlGaAs, $x = 0.30$ barriers. These wafers show a significant FWHM difference only for the narrow wells. While these wafers do not show the narrowest FWHM measured for the samples of this study, from these data we conclude that the lower growth rates used in MBE could be advantageous for the MOCVD growth of QW heterostructures having smooth interfaces. We have not yet tried growing QW's at 0.25 ml/s, but we expect that some benefit may be realized by lowering the growthrate further below the 0.5 ml/s rate employed in the present work.

The FWHM of the PL from the AlGaAs-GaAs QW's of this study are compared with data from the best values previously reported for similar structures grown by MBE, ALE, and FME in Fig. 8. The best MOCVD samples of the present study show FWHM values for thin wells ~6 ml that are comparable to the best 6 ml-thick QW's grown by MBE with 2 min interruptions. Except for the thicker wells, the MOCVD wells of Fig. 8 have PL lines much narrower that the similar continuously grown MBE QW's. The narrow QW's of this work show FWHM's that are better than previously reported for MOCVD-grown QW heterostructures.[4,5] Although earlier workers [15] have reported LP-MOCVD-grown QW's having a thickness of ~20 Å (~7 ml) with FWHM of ~6 meV at 2 K, an anomalous higher-energy peak was also present, indicating a possible fluctuation of 1 ml for this narrow well. The linewidths are also narrower than the FWHM of the PL QW's grown by ALE[7] and FME.[6]

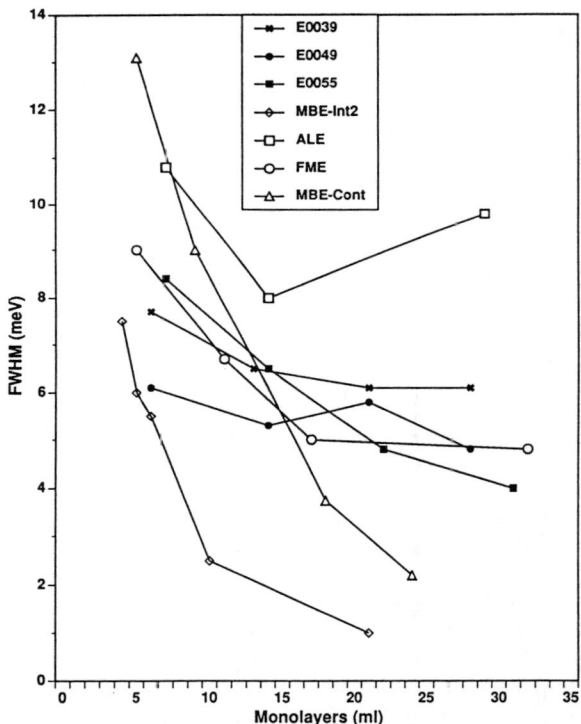

Fig. 8. FWHM of QW luminescence vs. QW thickness for various MQW wafers grown in the present work by MOCVD (closed crosses, squares, circles), and by ALE (open squares), FME (open circles), MBE with 120 s interruptions (open diamonds), and by continuous MBE (open triangles).

We believe that the FWHM of the PL emission from the thicker MOCVD QW's of this study are limited by the presence of impurities (both donors and acceptors) in the QW. It has been observed that the PL from thick quantum wells can be influenced by the emission from impurities.[14]

This presence of this extrinsic luminescence is consistent with the measured impurity concentrations in the undoped GaAs films grown with the present sources. The PL from the unintentionally doped GaAs layers grown under similar conditions is dominated by the presence of C acceptors. Weak PL is also observed from donor atoms but at a much lower concentration. Hall measurements on samples grown under other conditions (lower V/III ratios) shows p-type condition with a net hole concentration $\sim 10^{16}$ cm^{-3}. We expect that QW's grown with lower C concentrations will exhibit narrower PL emission for the thicker wells.

4. Summary

Low-pressure MOCVD has been used to grow AlGaAs-GaAs multiple-quantum-well heterostructures on GaAs substrates. The low-temperature (4.2 K) PL spectra from these MQW samples show that the QW's have smooth heterointerfaces on the order of the exciton diameter (~250 Å) over large distances ~2-8 mm. The FWHM is smallest for QW's grown in the 750–800 °C range and is not strongly affected by the growth interrupt time between layers. The growth rate dependence of the FWHM is not significant for the thicker wells (~15 ml) since the FWHM is probably dominated by the extrinsic component of the QW luminescence and not interface roughness. The QW PL is also not significant influenced by the interrupt times in the range 6-120 s used in the present study.

Acknowledgments

We thank Dr. Thomas Block, Robert Burnham, Steve Smith, and Richard Stall for useful discussions and Herbert Woodson and Ben Streetman for encouragement and support. The support of Amoco Technology Corporation, Epichem Ltd., Motorola, and United Technoligies is gratefully acknowledged. The work was partially supported by the National Science Foundation Science and Technology Center for the Synthesis, Growth, and Analysis of Electronic Materials.

References

[1] R.D. Dupuis, P.D. Dapkus, N. Holonyak, Jr., E.A. Rezek, and R. Chin, 1978 *Appl. Phys. Lett.* **32** 295
[2] C.W. Tu, R.C. Miller, B.A. Wilson, P.M. Petroff, T.D. Harris, R.F. Kopf, S.K. Sputz, and M.G. Lamont, 1987 *J. Crystal Growth* **81** 159
[3] R.P. Kopf, E.F. Schubert, T.D. Harris, and R.S. Becker, 1991 *Appl. Phys. Lett.* **58** 631
[4] H. Kawai, K. Kaneko, and N. Watanabe, 1984 *J. Appl. Phys.* **56** 463
[5] N. Watanabe and Y. Mori, 1986 *Surface Sci.* **174** 10
[6] N. Kobayashi, T. Makimoto, Y. Yamauchi, and Y. Horikoshi, 1989 *J. Appl. Phys.* **66** 640
[7] J.R. Gong, P.C. Colter, D. Jung, S.A. Hussein, C.A. Parker, A. Dip, F. Hyuga, W.M. Duncan, and S.M. Badair, 1991 *J.Crystal Growth* **107** 83
[8] S.B. Ogale, A. Madhukar, F. Voillot, M. Thomsen, W.C. Tang, T.C. Lee, J.Y. Kim, and P. Chen, 1987 *Phys. Rev.B* **36** 1622
[9] R.C. Miller, C.W. Tu, S.K. Sputz, and R.F. Kopf, 1986 *Appl. Phys. Lett* **49** 1245
[10] W.T. Tsang, F.K. Reinhart, and J.A. Ditzenberger, 1980 *Appl. Phys. Lett* **36** 118
[11] T. Fukunaga, K.L.I. Kobayashi, and H. Nakashima, 1985 *Japan J. Appl. Phys.* **24** L510
[12] C.A. Warwick, W.Y. Jan, A. Ourmazd, and T.D. Harris, 1990 *Appl. Phys. Lett* **56** 2666
[13] F. Voillot, A. Madhukar, J.Y. Kim, P. Chen, N.M. Cho, W.C. Tang, and O.G. Newman, 1986 *Appl.Phys.Lett.* **48** 1009
[14] R.C. Miller, A.C. Gossard, W.T. Tsang, and O. Munteanu, 1982 *Phys. Rev. B* **25** 3871
[15] M.-E. Pistol, S. Nillson, P. Silverberg, L. Samuelsom, M. Rask, and G. Landgren 1986 *Superlattice and Microstructures* **2** 501

Joint Soviet-American Workshop on the Physics of Semiconductor Lasers May 20–June 3 1991

In situ patterning of impurity induced layer disordering and other applications of laser patterned desorption

J.E. Epler[a]
Paul Scherrer Institute, Badenerstrasse 569, CH–8048 Zürich, Switzerland
R.D. Bringans, T.L. Paoli, and D.W. Treat
Xerox Palo Alto Research Center
3333 Coyote Hill Road, Palo Alto, CA 94304 USA

Abstract. Laser patterned desorption (LPD) is an *in situ* processing technique that enables partial or complete removal of quantum well (QW) layers within GaAs-AlGaAs device heterostructures. In LPD, the evaporation rate of GaAs and incorporated dopants is greatly enhanced by local heating with Ar^+ and Nd:YAG laser beams within a metalorganic chemical vapor deposition reactor. Typically, a small area (8 μm to 3 mm in diameter) of the GaAs QW layer is exposed during a pause in the epitaxial crystal growth. After exposure, growth is resumed, burying the patterned layer(s) within the crystal. Various techniques are used to characterize the effect of the LPD process and several applications of LPD are shown. The most recent application is the patterning of the impurity induced layer disordering of a GaAs-AlGaAs superlattice by locally removing Si doping spikes.

Introduction of patterning techniques into the epitaxial growth system is an essential part of developing a true three-dimensional heterostructure capability. Although a large number of *in situ* processing techniques have been described,[1] the number of device demonstrations based upon combined *in situ* processing and/or epitaxial growth are considerably fewer. Examples of the latter include: a quantum well (QW) diode laser with a laser written active region,[2] multiple wavelength light emitting diodes[3] by laser-enhanced growth and a multiple wavelength laser array[4] by laser patterned desorption (LPD). In this paper we review two applications of LPD: grading of a quantum well active layer[4] and patterning the current injection into a laser diode bar,[5] and introduce a third: *in situ* patterning of the impurity induced layer disordering (IILD) process.

The essence of LPD is the use of laser heating to increase the evaporation rate of a GaAs surface during a pause in the epitaxial growth. As previously described,[4,5] GaAs-AlGaAs heterostructures are grown by metalorganic chemical vapor deposition (MOCVD) in an inverted chimney chamber operated at atmospheric pressure. In this system, the GaAs substrate is horizontal, facing downward and the bottom of the reactor chamber is a quartz window through which CW Ar^+ and Nd:YAG laser beams are introduced. For the first application, the Ar^+ laser is operated in the TEM_{00} mode, single line at 514.5 nm with a power output of 2.4 W. The Nd:YAG beam is multi-mode with a 5.0 W output. The lasers are aligned to coincide on the surface of the sample to provide maximum increase in the surface temperature.

The epitaxial layers are grown at the optimum temperature of 800 °C and include (in order of growth) a Se-doped GaAs buffer layer (0.3 μm), a Se-doped $Al_{0.8}Ga_{0.2}As$ lower confining layer (1.0 μm), an undoped $Al_{0.4}Ga_{0.6}As$ waveguide layer (0.6 μm), and a GaAs quantum well (13.6 nm). At this point in the growth, the metalorganic sources are vented and a 1% arsine-hydrogen mixture is introduced into the chamber. The substrate temperature is increased to 825 °C and the combined laser beams are applied to the surface. For 90 s, GaAs is desorbed from the surface at a greatly enhanced rate within the laser heated spot. The temperature at the spot center is approximately 1000 °C. After the desired desorption is completed, the substrate temperature is returned to 800 °C and growth is resumed. The remaining layers are another $Al_{0.4}Ga_{0.6}As$ waveguide layer (0.6 μm), a Mg-doped $Al_{0.8}Ga_{0.2}As$ upper confining layer (0.9 μm), and a Mg-doped GaAs cap layer.

The transmission electron microscope (TEM) cross sections of the waveguide regions at four different positions relative to the center of the 3-mm diameter laser spot are shown in Fig. 1. These TEM cross sections are taken using the corner incidence method[6] and provide an accurate determination of the thickness of the crystal layers. In Fig. 1(a), near the center of the laser spot, the GaAs QW has been thinned to 7.4 nm during the 90 s exposure. The 6.2 nm of GaAs has been desorbed at a rate of 1 monolayer per 4 s. Consistent with the thermal profile, the QW thickness gradually increases to (b) 8.5 nm at 0.7 mm from spot center, (c) 11.3 nm at 1.3 mm distance, and (d) 13.6 nm at 2.7 mm distance. The 13.6 nm QW is representative of the region unilluminated by the lasers.

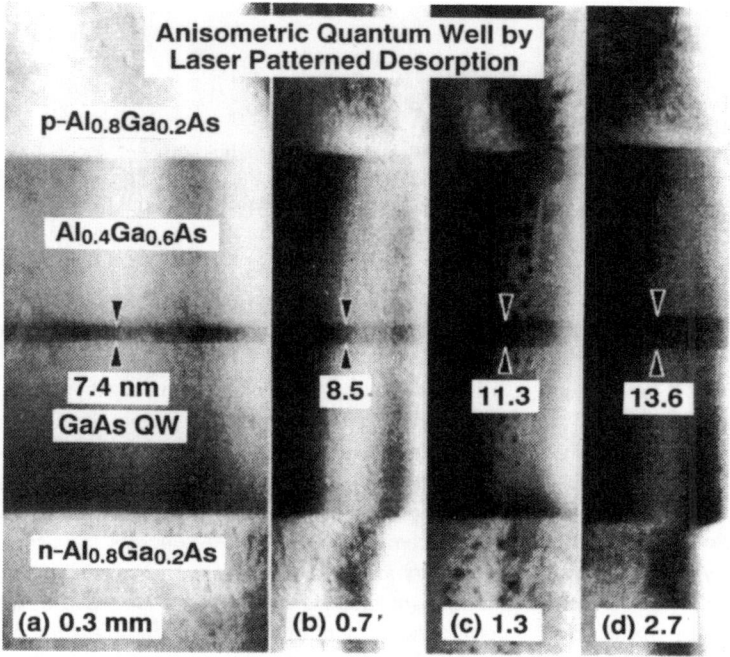

Fig. 1. TEM cross sections at four positions along the laser desorbed region of the sample. The quantum well thickness and distance from the spot center are respectively: (a) 7.4 nm, 0.3 mm; (b) 8.5 nm, 0.7 mm; (c) 11.3 nm, 1.3 mm; and (d) 13.6 nm, 2.7 mm.

For a device demonstration, the wafer is processed in standard fashion into broad area laser bars. The bars are 1 cm long with shallow saw cuts to electrically isolate neighboring devices.

Each individual broad area laser of the bar is 250 μm long and 250 μm wide. The threshold current and emission wavelength as a function of bar position are shown in Fig. 2. The data points are shown for every other laser diode. Including the intermediate devices, several individually addressable wavelengths can be obtained from each half of the bar. The emission wavelength varies from 792 nm in the center of the laser spot to 830 nm in the field, with the graded regions lasing at intermediate wavelengths. The threshold current density varies from 500 A/cm² in the field to 480 A/cm² in the center of the laser spot. The threshold current density is lower within the center of the laser spot because the 13.6 nm QW in the field is thicker than optimum.

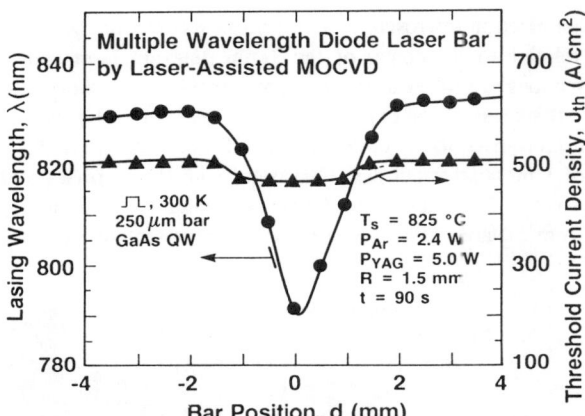

Fig. 2. The lasing wavelength and threshold current density as a function of position along the wafer. The decrease in the lasing wavelength corresponds to the thinning of the quantum well.

Arrays of high power lasers have been fabricated from laser-desorbed material. Shallow proton bombardment through a mask of thick photoresist is used to localize current on the p-side to form stripe-geometry, gain-guided lasers. Four individually addressable lasers each consisting of ten stripes on 10-μm centers have been fabricated on a bar 2 mm long by 250 μm wide. One facet on each bar is coated with a highly reflective dielectric stack, while the other facet is coated with a half-wave of alumina. Each chip is mounted junction-side up to a copper heatsink. Thus only pulsed operation is possible.

The four emission spectra of a multiple wavelength array are given in Fig. 3. At 1.2 times threshold, the center wavelength varies from 806 nm to 835 nm across the chip. Outputs from adjacent arrays were spectrally separated by a minimum of 8 nm to a maximum of 15 nm across a total bandwidth of 30 nm. The multiple longitudinal mode spectra is consistent with normal 10-stripe laser array operation. The corresponding light output versus current of each array element is shown in Fig. 4. The thresholds varies from 90 mA for the shortest wavelength array to a maximum of 120 mA for the array operating at the longest wavelength. Differential efficiencies were uniformly 35% while the maximum power output at each wavelength is well over 200 mW. Consistent with the above results, the threshold current density is lowest for device fabricated from the center of the laser exposed region. The broad area threshold current density of 500 A/cm² for these arrays compares favorably with commercially available 10-stripe laser arrays. Applications for this type of device could include a wavelength multiplexing communication source.

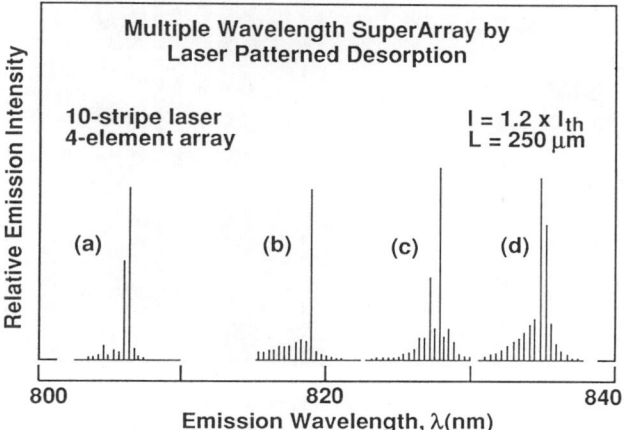

Fig. 3. Emission spectra of the four element array. The spectra are typical of a gain-guided 10-stripe laser. The threshold currents are (a) 90 mA, (b) 95 mA, (c) 100 mA, and (d) 120 mA.

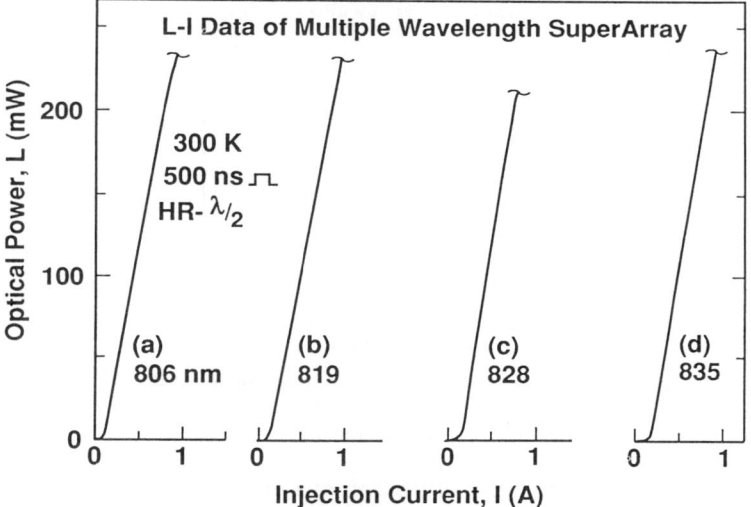

Fig. 4. Optical power versus injection current (L–I) for each subarray of a four-wavelength SuperArray. The devices are limited to pulsed operation because of the junction-side-up bonding.

As a second topic, LPD is used to completely remove a reverse biased GaAs QW which serves as a current blocking layer in a laser diode structure. The goal is to control the current injection into a laser diode bar and assess the feasibility of defining striped active regions. The device structure can be seen in the scanning electron microscope (SEM) cross section of Fig. 5(c) and is similar to that of the multiple wavelength laser except that the heavily Se-doped QW of 7.5 nm thickness is added to the middle of the p-doped AlGaAs cladding layer. After growth of the reverse biased layer, a 3-mm wide region is exposed for 300 s. The other conditions are identical to those

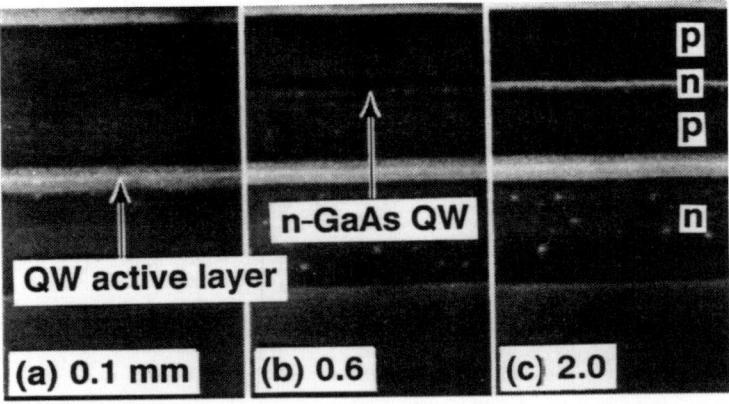

Fig. 5. SEM cross section of three regions with respect to the laser spot center. In (a) the current blocking GaAs QW is totally removed, in (b) a discontinuous remnant remains, and in (c) the QW is undesorbed.

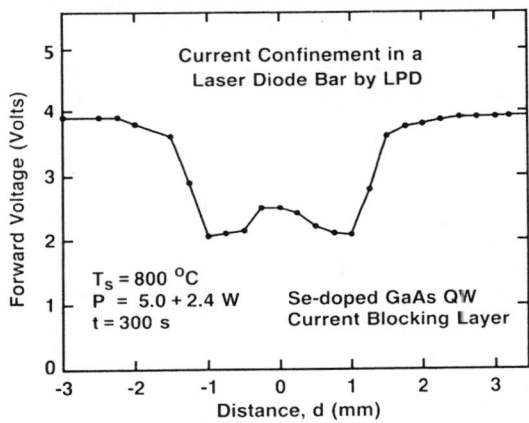

Fig. 6. Forward voltage (at 10 mA) as a function of position along the wafer. The 3-mm extent of the desorbed region corresponds to the spot size of the laser radiation.

described above. In Fig. 5(a), near the center of the laser spot, the GaAs QW has been removed during the 300 s exposure. (TEM analysis of similar experiments has confirmed this assertion). Only a very faint trace of the layer is visible in the original micrograph. Since the A-B etch is very sensitive to crystalline defects, we conclude that no gross damage has occurred at the interface. In Fig. 5(b), 0.6 mm from the laser spot center, but still well within the laser exposed region, some of the QW layer remains as shown by the discontinuous thin line, the remnant of the QW. At 2.0 mm from the laser spot center (c), the reverse-biased layer is clearly seen. This location is representative of the field outside the laser spot illumination where the p-n-p-n structure provides an effective barrier to current injection.

To demonstrate the application of laser-patterned blocking layers, the wafer is processed into 250 μm wide diode laser bars. As previously reported,[5] shallow saw cuts are made to electrically isolate 250 μm long by 250 μm wide broad area lasers. The forward voltage (at a current density of 16 A/cm^2) is shown as a function of bar position in Fig. 6. The spatial variation of the forward voltage clearly indicates the region over which the reverse-biased layer has been removed. Near the center of the laser spot the forward voltage increases to 2.5 V, probably as a result of overexposure of the AlGaAs. However the broad area devices in and out of the exposed area generally exhibit low threshold currents (200 to 280 mA) and no other evidence of crystal damage is observed. Current versus voltage curves and observation of the near field of the electroluminescence[5] all support the conclusions that the current in a laser bar can be channelled into a selected region. The scale of this experiment has been successfully reduced to 40 μm; however, further optimization of the growth and exposure conditions are necessary to achieve the goal of a narrow current injection stripe with low leakage current in the surrounding area.

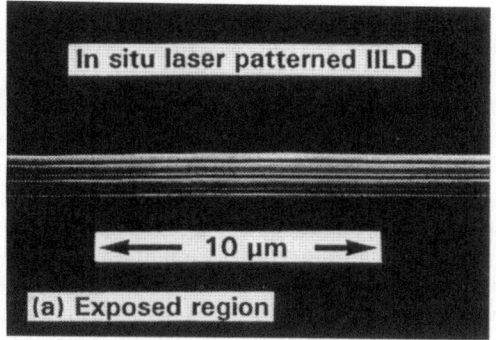

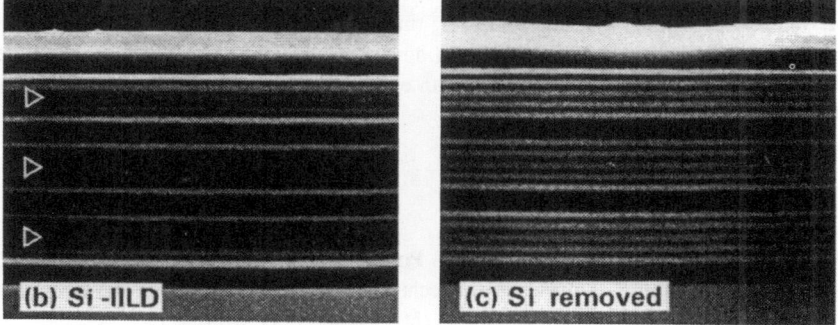

Fig. 7. (a) SEM cross section of the region of the epitaxial layers where the Si spikes have been at least partially removed. The 8-μm extent of the non-disordered SL corresponds to the laser spot size; (b) unexposed part of the SL disordered by Si-IILD; (c) SL section where IILD has been inhibited by removal of Si doping.

The third area of application is the patterning of the IILD process within the MOCVD reactor. IILD has been widely used to fabricate high quality buried heterostructure lasers and waveguides[7] and would be a powerful tool if successfully integrated with the crystal growth. The central concept of IILD is that heterostructures are naturally unstable in the presence of large concentrations of mobile point defects such as vacancies or interstitials. As discussed earlier,[7] the concentration of point defects is a strong function of the position of the Fermi level. For example, the concentration of column III vacancies in a GaAs-AlGaAs heterostructure is greatly increased in the presence of heavy Si doping ($> 10^{18}$ cm^{-3}). The challenge for using LPD in combination with Si-IILD is to locally remove the Si impurity by enhanced evaporation.

For this application the structure best depicted in Fig. 7(c) is grown. An undoped GaAs buffer layer is followed by a 180-nm $Al_xGa_{1-x}As$ ($x = 0.4$) marking layer. Then 3.5 cycles of a GaAs-$Al_xGa_{1-x}As$ ($x = 0.4$) superlattice (SL) of 84 nm period are grown. After the end of the third GaAs layer (arrow) a Si doping spike is deposited by flowing a silane-hydrogen-arsine mixture for 2 s. Then an extremely thin (a few monolayers) of GaAs is deposited. The growth is paused and the Ar^+ laser beam (approximately 200 mW, 8 μm spot size) is slowly scanned across the surface (50 $\mu m \cdot s^{-1}$) using galvanometric controlled mirrors. The growth is then resumed and the process of depositing and then removing a Si doping spike is repeated twice more at the layers indicated with an arrow. The structure is completed with a fourth marking layer and a GaAs cap. Finally, the sample is annealed in the reactor at 850 °C for 15 min to provide time for the Si to diffuse, intermixing the Ga-Al sublattice. In Fig. 7(a), a low resolution SEM micrograph shows the lateral extent of the laser inhibited IILD without showing the details of what actually occurred. In Fig. 7(b), at a 6 times higher magnification the partially disordered SL is shown. Note that the SL layers closest to the doping spike are most disordered. For comparison, in (c) is given the cross section of a region in the laser spot center. The contrast is unmistakable and indicates that the LPD is removing some if not all of the Si dopant. Further work is required to assess the applicability of this process to device fabrication.

In summary, three application of LPD have been discussed in connection with the three dimensional patterning of GaAs-AlGaAs heterostructures. Each application indicates a different area of device processing that can be attempted *in situ*. The graded quantum well has applications for window lasers or single-growth distributed bragg reflector lasers where transparent sections of waveguide are desired. The current confinement technique may prove useful in stripe laser definition or electrical isolation in general. And finally, the patterning of IILD would enable buried heterostructure laser fabrication without the long annealing times generally associated with this process. The future of any of these techniques in commercial device production is difficult to predict. However the potential of integrating lateral patterning with epitaxial crystal growth is tremendous and LPD represents one such approach.

Acknowledgments

The authors wish to thank G. Anderson and F. Ponce for the TEM micrographs and H. Chung, S. Nelson, F. Endicott, and R. Donaldson for technical assistance. This work has been supported in part by the Defense Advanced Research Projects Agency (J.D. Murphy).

References

[a] Work performed at Xerox Palo Alto Research Center
[1] D.J. Ehrlich, J.G. Black, M. Rothschild, and S.W. Pang 1988 *J. Vac. Sci. Technol.* **B6** (3) 895–899
[2] Q. Chen, J.S. Osinski, and P.D. Dapkus 1990 *Appl. Phys. Lett.* **57** 1437–1439
[3] J.E. Epler, H.F. Chung, D.W. Treat, and T.L. Paoli 1988 *Appl. Phys. Lett* **52** 1499–1501
[4] J.E. Epler, D.W. Treat, S.E. Nelson, and T.L. Paoli 1990 *IEEE J. Quantum Electron.* **26** 663–668
[5] J.E. Epler, D.W. Treat, and T.L. Paoli 1990 *Appl. Phys. Lett.* **56** 1828–1830
[6] H. Kakibayashi and F. Nagata 1985 *Japan J. Appl. Phys.* **24** L905–L907
[7] D.G. Deppe and N. Holonyak 1989 *J. Appl. Phys.* **64** R93–R113

Neutral impurity disordering of III–V quantum well structures for optoelectronics

J.H. Marsh, S.G. Ayling, A.C. Bryce, S.I. Hansen, and S.A Bradshaw

Department of Electronics and Electrical Engineering, The University,
Glasgow G12 8QQ, Scotland

Abstract. Novel applications of impurity induced disordering (IID) in semiconductor integrated optoelectronics are discussed and some requirements of the IID process are quantified. The effect of the neutral impurities boron and fluorine as disordering species has been studied. In the GaAs/AlGaAs system fluorine disordered multiple quantum well waveguide structures exhibited blue shifts of up to 100 meV in the absorption edge (representing complete disordering). The absorption coefficient in partially disordered structures at near band-edge wavelengths was as low as 4.7 dB cm^{-1}. This absorption edge shift was accompanied by substantial changes, > 1%, in the refractive index. Disordering of GaInAs/AlGaInAs and GaInAs/GaInAsP quantum well structures lattice matched to InP has also been investigated. The temperature stability of as-grown P-quaternary material is poor with blue shifts of the exciton peak occuring at temperatures greater than 500 °C but the Al-quaternary is stable to at least 650 °C. Boron implantation caused small (10 meV) blue shifts of the exciton peak in both material systems at low annealing temperatures. Much larger blue shifts (up to 90 meV for phosphorus quaternary and 45 meV for aluminium quaternary samples) were observed in the fluorine implanted samples.

1. Introduction

Impurity induced disordering (IID) is emerging as a powerful process for use in the fabrication of photonic and optoelectronic integrated circuits.[1] During the IID process both the bandgap and refractive index of a quantum well (QW) structure are modified, thus providing a route to alter the material absorption edge and to form optical waveguides. The physics of media containing multiple layers of quantum dimensions and of the disordering process itself are, however, complex and it is important at the outset to define clear objectives for the IID process in specific integrated optics applications. Here we quantify some objectives and present results on bandgap changes, absorption coefficient changes and refractive index changes induced by the neutral impurities fluorine and boron. The bulk of the results refer to the GaAs/AlGaAs material system, but some results refer to the GaInAs/AlGaInAs and GaInAs/GaInAsP systems with a bandgap wavelength of 1.5 μm.

Impurity induced disordering is a technique for increasing the bandgap of QW structures after growth by intermixing the wells with the barriers to form an alloy semiconductor with a bandgap larger than the original QW. In the process an impurity is introduced into the wafer either by diffusion or by ion-implantation and the wafer is then annealed. During the annealing step the layers intermix and ion-implantation damage, if present, is to a large extent removed. Current understanding[2] of the IID process suggests that the role of impurities is to induce the disordering process through the generation of free-carriers which, in turn, increase the equilibrium number of vacancies at the annealing temperature. A number of species has been demonstrated to disorder the GaAs/AlGaAs system, the most important of which are Zn (p-type) and Si (n-type). Impurities need to be present in a concentration greater than around 10^{18} cm^{-3} in order to enhance the interdiffusion rates of the lattice elements. In the case of impurities which are active dopants at room temperature, this gives rise to high free-carrier absorption losses in waveguides, greater than 10 cm^{-1}. We, however, have made significant advances in QW disordering in the GaAs/AlGaAs material system by using boron and fluorine, which are neutral at room temperature, and we have demonstrated that total propagation losses as low as 4.7 dB cm^{-1} (1.1 cm^{-1}) can be realised[3] using fluorine implantation together with bandgap increases[4] of up to 90 meV.

QW structures exhibit a number of polarisation sensitive effects, most significantly a polarisation dependent dichroism and birefringence but the polarisation dependences disappear as the structures are intermixed and become more like bulk alloys. The dichroic effect arises from the selection rules governing optical absorption in a QW with the TE polarisation exciting transitions from both the heavy hole (HH) and light hole (LH) confined states into the conduction band states, and the TM polarisation exciting transitions only from the LH states. The absorption edge therefore occurs at a longer wavelength for the TE polarisation than for the TM polarisation. Birefringence arises because QW structures consist of a number of dielectric layers, each layer being much thinner than the wavelength of light, and the effective dielectric constant of the composite structures therefore depends on whether the optical electrical field is parallel or perpendicular to the plane of the wells.[5] However, for wavelengths close to the absorption edge, the refractive index spectrum is determined to a substantial extent by the rapidly changing absorption spectrum. Because the absorption spectrum is strongly anisotropic the birefringence increases markedly as the absorption edge is approached—we have recently measured this directly in partially disordered GaAs/AlGaAs MQW waveguide structures[6] as summarised below.

Only a limited amount of work has been carried out in longer wavelength materials: in the lattice matched GaInAs/InP system disordering using both sulphur[7] and high concentration proton[8] implants have been demonstrated to give increases in the bandgap energy (intermixing both the group III and group V lattice sites) whilst zinc[9-11] gives bandgap reductions (intermixing only the group III lattice sites). The amphoteric impurities, germanium[12] and silicon,[13] and the isoelectronic impurities, gallium[14] and phosphorus,[15] have also been demonstrated to give bandgap increases if implanted using high doses. The use of AlInAs as an alternative to InP and AlGaInAs as an alternative to GaInAsP means that only the group III sites need to be intermixed. The effectiveness of fluorine and boron in disordering this system will be discussed and compared to the GaInAs/GaInAsP system.

2. Applications of impurity induced disordering in components for coherent systems

A number of potential applications of the IID technique in integrated optoelectronics for coherent applications can be identified:

2.1. Low loss waveguides for interconnecting components on an OEIC (Fig. 1(a))

Two parameters are of particular importance in this application: the absorption coefficient (α) and the material resistivity (ρ). An ideal target for α is $\alpha < 1$ dB cm^{-1} but 10 dB cm^{-1} would be acceptable in many applications. A problem arises when active dopants are used as disordering species: the threshold concentration of impurities necessary to induce the IID process is typically

$> 10^{18}\,\text{cm}^{-3}$. The most commonly reported impurity in IID is Si and the lowest reported absorption coefficients are around 43 dB cm^{-1} (10 cm^{-1}) which is a consequence of free carrier absorption. A further requirement is that the electrical resistance of waveguides should be sufficiently high to isolate individual components. Studies of an integrated laser/modulator structure[16] have demonstrated that the required isolation resistance between the laser and the modulator, R, is $R \geq 100\,\text{k}\Omega$. For waveguide dimensions of $3 \times 1\,\mu\text{m}^2 \times 0.5$ mm long, a carrier density below $10^{17}\,\text{cm}^{-3}$ is therefore needed. There is clearly a trade-off between the required electrical isolation and the tolerable optical attenuation in designing a waveguide for interconnection, but it appears that Si (or Zn) IID is unlikely to give the required performance. As a consequence, the studies reported here have used the electrically neutral dopants F and B.

2.2. Integrated passive waveguides for line-narrowed lasers (Fig. 1(b))

Semiconductor lasers are usually line-narrowed by operating them in external cavities which are both bulky and subject to alignment problems. Integrated cavities are mechanically stable and will be around a factor of four shorter than the equivalent air cavity. Estimates of the linewidth reduction can be made based on the Schawlow–Townes formula for the laser linewidth:

$$(\Delta v)_{laser} = \frac{2\pi h v (\Delta v_c)^2 n_{spon}}{P_{out}}(1+\alpha_H^2) + A$$

where the cavity linewidth:

$$\Delta v_c = \frac{1}{2\pi}\frac{c}{n}\left[\frac{\alpha_a L_a + \alpha_e L_e}{L_a + L_e} - \frac{1}{L_a + L_e}\ln(R_1 R_2)^{1/2}\right], \quad (1)$$

α_a and L_a are the absorption coefficients and length of the active region, α_e and L_e are the absorption coefficients and length of the passive cavity region, n_{spon} is the spontaneous emission factor, and α_H is the linewidth enhancement factor. In a QW laser the gain coefficient at threshold is given by:[17]

$$G_{th} = N\Gamma_w \gamma_0 \ln\left[\frac{J_{th}}{N J_T}\right] \quad (2)$$

where N is the number of wells, Γ_w is the overlap between the optical wave and a single well, J_{th} is the threshold current density, and $\gamma_0 = 690$ cm^{-1} and $J_T = 66$ A cm^{-2} are experimental values for 100 Å wells in GaAs.

Tab. 1. The estimated reduction in lasing linewidth for integrating a 300 μm active region with passive cavities of lengths.

Propagation loss in passive region, α_e (cm^{-1})	1	2	5	10
Maximum length of passive region, L_e (cm)	1.3	0.65	0.26	0.13
Linewidth reduction factor	542	142	26	7.8

Our double QW lasers have a threshold current density of 420 A cm^{-2} (consistent with the values of γ_0 and J_T given above), and a propagation loss (arising principally from free-carrier absorption from the carriers injected into the active region) of 10 cm^{-1} when lasing. Using these parameters Tab. 1 shows the estimated reduction in linewidth for an integrated extended cavity laser with an active region of length 300 μm operating with a threshold current three times that of a non-extended cavity laser. This value of threshold current density is similar to that of a double heterostructure (i.e. non-QW) laser. In practice larger reductions than those given in Tab. 1 would be expected because the linewidth enhancement of an extended cavity laser is effectively reduced.

2.3. Non-absorbing mirrors (Fig. 1(c))

The power output from GaAs/AlGaAs lasers is limited by facet degradation which occurs as a result of optical absorption in the regions adjacent to the facets. This effect can be reduced by widening the bandgap at the facets. In this application it is desirable to keep carriers away from the facets: this implies a bandgap increase of several kT is required.

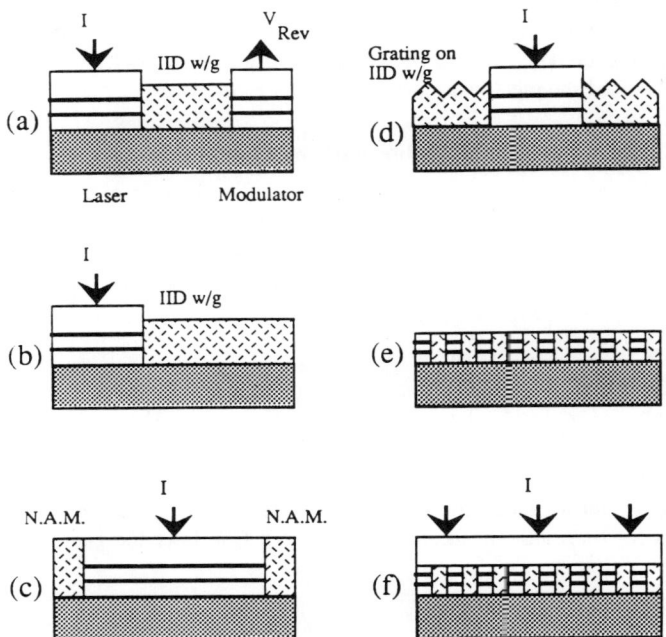

Fig. 1. Examples of the potential applications of IID in integrated optoelectronics: (a) integrated laser and modulator, (b) extended cavity narrow linewidth laser, (c) high power laser with non-absorbing mirrors, (d) DBR laser, (e) grating formed by IID and (f) DFB laser with index and gain gratings.

2.4. Single frequency DBR lasers (Fig. 1(d))

Conventional DBR lasers either suffer from high optical attenuation in the grating regions or require a complex series of overgrowth processes. It is in principle possible to form IID waveguides at either end of a QW laser and etch gratings onto these waveguides. In order for the laser to be single mode it can been shown[18] that $kl_{cav} \leq 3$ where k is the coupling coefficient and l_{cav} is the effective length of the laser cavity. A grating of length l_{gr} with an absorption coefficient α needs $kl_{gr} \approx 1$ and $\alpha \ll k$ (although $\alpha < k/2$ is not unreasonably high). Hence, for a laser cavity 300 μm long, $l_{gr} \approx 100\,\mu m$, $k \approx 100\,cm^{-1}$ and $\alpha < 50 cm^{-1}$ (220 dB cm^{-1}). Absorption coefficients as high as this would not be practical as half the output power would be absorbed in the facet, but values of α at least a factor of 5 lower than this are achievable even through active dopant (Si, Zn etc) IID. Low coupling coefficient gratings could be formed on extended cavity lasers (Sec. 2.2) to give stable narrow-linewidth sources.

2.5. Gratings formed through IID induced changes in the refractive index (Fig. 1(e))

The dichroic and birefringent properties of QW waveguides are modified (and the limiting case removed) by the disordering process and the spectral distributions of the absorption coefficient and of the refractive index are changed. It is possible to use the refractive changes to form gratings directly,[19] rather than, for example, by etching an air grating, but accurate values of the refractive index of both disordered and non-disordered material are needed.

2.6. DFB gain and phase gratings (Fig. 1(f))

A further application of the IID technique is in DFB lasers. In conventional DFB lasers a grating is directly etched into a region close to the active layer, and the structure is then overgrown with the upper cladding layer and contact layer. This requires a structured surface to be overgrown and it is generally more desirable to overgrow planar surfaces which could be achieved by defining gratings in a superlattice cladding region by IID. Most exciting is the prospect of implanting and diffusing into the active wells themselves since, if the IID material is of higher resistivity than the undoped material, current will be preferentially injected into the non-disordered regions. This would result in a simultaneous gain and phase grating which could be designed to concentrate the gain at the points of maximum optical intensity longitudinally along the cavity: a considerable reduction in the threshold current density would also be expected. This is an intermediate step towards the fabrication of quantum wire or dot lasers. In applications of these types accurate values of bandgap steps, resistivity and refractive index are all required.

3. IID of GaAs/AlGaAs

3.1. Sample processing

Two types of GaAs/AlGaAs structure were studied: multiple QW (MQW) and double QW (DQW). Growth in both cases was by atmospheric pressure metal-organic vapour phase epitaxy (MOVPE).[20] Boron or fluorine ions were introduced by implantation, and subsequent disordering took place during an annealing stage in a conventional furnace with the samples capped by a dielectric layer. In this section details of the sample design and individual processing steps are described.

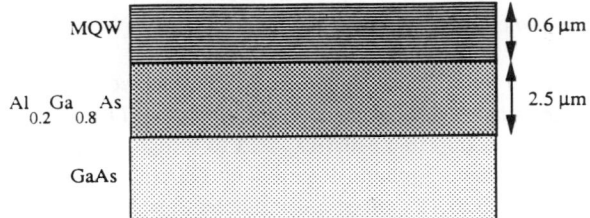

Fig. 2. The multiple quantum well (MQW) waveguide structure used for the IID studies. The MQW consisted of 37 periods of 80 Å GaAs wells and 80 Å $Al_{0.26}Ga_{0.74}As$ barriers.

The MQW structure (Fig. 2) was designed to be suitable both for analysis by photoluminescence spectroscopy (PL) and for waveguide studies. An MQW high-index region consisting of 44 GaAs wells and $Al_{0.26}Ga_{0.74}As$ barriers, both with a nominal thickness of 8 nm was therefore grown at the top of the structure. Optical confinement was provided by air above the core region and by a thick $Al_{0.20}Ga_{0.80}As$ lower cladding layer, thus forming a single vertical mode at the wavelength of interest. The detailed parameters of the structure were intended to ensure that only the lowest order depth mode would propagate in the waveguide experiments.

The MQW structure is not suitable for optoelectronic integration involving lasers and other similar devices. We have therefore also examined the disordering of DQW waveguide structures compatible with laser fabrication (but at this stage undoped to eliminate possible complications in the impurity-disordering process associated with grown-in dopants). The DQW structure (Fig. 3) consisted of a semi-insulating GaAs substrate on which were grown, in sequence, a GaAs buffer layer 0.2 μm thick, a 2.0 μm thick $Al_{0.4}Ga_{0.6}As$ buffer layer, a waveguide core region consisting of two GaAs wells 10 nm thick with an $Al_{0.2}Ga_{0.8}As$ barrier layer 10 nm thick between the wells and barriers 0.1 μm thick on either side of the wells. A top cladding layer of $Al_{0.4}Ga_{0.6}As$, 1.0 μm thick, and a thin layer of GaAs completed the structure.

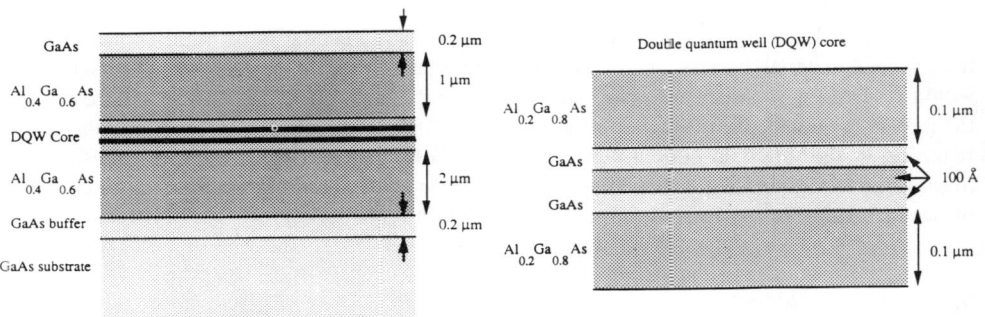

Fig. 3. The double quantum well (DQW) epitaxial waveguide structure used in the IID studies.

Samples typically of area about 1 cm² were implanted with either boron or fluorine ions with the aim of producing a concentration in the QW region between 3×10^{16} cm^{-3} and 3×10^{19} cm^{-3}. Due to the localized nature of the impurity implantation, implantation at three separate energies was generally used for the MQW samples with aggregate doses of between 3×10^{12} cm^{-2} and 3×10^{15} cm^{-2}. For a total dose of 10^{14} cm^{-2}, for example, using fluorine, the implantation doses and energies were: 1.53×10^{13} cm^{-2} at 80 keV; 3.3×10^{13} cm^{-2} at 260 keV; 5.1×10^{13} cm^{-2} at 700 keV, and using boron: 1.8×10^{13} cm^{-2} at 50 keV; 3.3×10^{13} cm^{-2} at 170 keV; 4.9×10^{13} cm^{-2} at 400 keV. The DQW sample was implanted with fluorine with a dose of 6×10^{12} cm^{-2} at a single energy of 1 MeV.

Annealing was carried out in a conventional diffusion furnace with a high-purity flowing nitrogen atmosphere. The samples were mounted inside an enclosed high-purity graphite box so that the sample top surfaces were uppermost and exposed to a high local vapour pressure of arsenic provided from a small volume of gallium loaded with GaAs. All samples were capped with a plasma-deposited layer of 100 to 150 nm of either silicon dioxide or silicon nitride. It was found that the two types of capping layer gave very similar results and so SiO_2 capping layers were the main ones used because of the relative ease with which they could be removed if required. It was also concluded that, under our deposition and annealing conditions, any contribution to intermixing from vacancies created by diffusion of gallium into the SiO_2 capping layer[21] was relatively small. Annealing was investigated over the range of temperatures from 750 °C to 920 °C (measured with a thermocouple embedded in the graphite box). Unimplanted, but capped, control samples were placed alongside the implanted samples to allow both the relative and absolute shifts of sample properties caused by the implanted impurities to be determined.

3.2. Photoluminescence measurements

Photoluminescence spectroscopy (at 18 K) was used to optimise the implantation and annealing conditions.[4] Features associated with recombination at the bandgap and those associated with damage were identified and compared in intensity: from these measurements the optimum implant dose was found to be around 10^{18} cm^{-3} (10^{14} cm^{-2}) with an annealing temperature of 890 °C; however at higher implant doses (10^{15} cm^{-2} range) or for longer anneals at lower temperatures the peaks associated with damage became broader and more intense. Doses of 3×10^{17} cm^{-3} (3×10^{13} cm^{-2}) and lower were found to progressively less effective at inducing disordering. The use of annealing temperatures above 910 °C resulted in significant bandgap increases in the control samples.

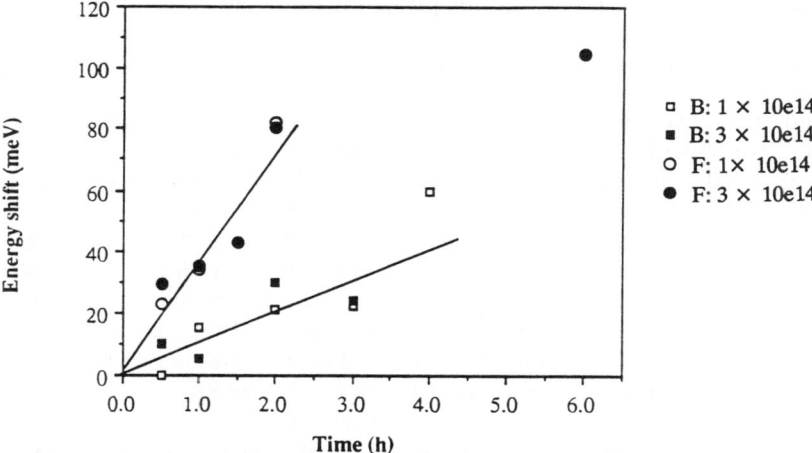

Fig. 4. Bandgap increases associated with boron and fluorine IID using an annealing temperature of 890 °C. The aggregate implant doses per cm^2 are indicated.

Fig. 4 shows the variation of the energy shift of the bandgap with annealing time at 890 °C, for two different fluorine and boron implantation doses. Using fluorine the energy shift, at times for which the mixing process does not approach saturation, is over twice that observed using boron. The required annealing conditions (2 h at 890 °C results in a bandgap increase of > 60 meV using F) appear to be compatible with retaining good device performance from lasers and modulators: we have subjected a double QW laser wafer to an anneal cycle at 905 °C for 2 h and then fabricated the wafer into broad area lasers. The threshold current density after annealing was 415 A cm^{-2}, the same (to within 5%) as that measured on non-annealed lasers.

3.3. Propagation loss measurements

Room temperature absorption measurements were carried out on rib waveguides and the losses of unimplanted, unannealed control samples, fluorine IID samples and boron IID samples compared. The IID samples were implanted with impurity doses of 10^{18} cm^{-3} and annealed at 890 °C for 2 and 4 h respectively. Rib waveguides were defined by dry-etching (using SiCl$_4$) completely through the 0.7 μm deep disordered MQW layer to leave waveguide ridges 4 μm wide which were multi-moded laterally and single-moded vertically. Light was end-fire coupled using 40X microscope objectives into and out of the waveguides. Initial estimates of the propagation loss were made using relatively short (2–3 mm) waveguides and measuring the optical power into and out from the end-fire rig. Assuming a coupling loss of 3 dB, a value which accounts for Fresnel reflection losses at the waveguide facets but makes no allowance for modal mismatch, an upper estimate of the propagation loss can be made. The propagation losses, measured at a wavelength of 885 nm, were 7 dB cm^{-1} in the starting material, 13 dB cm^{-1} in fluorine-disordered material and 22 dB cm^{-1} in boron-disordered material. The additional losses introduced by the fluorine and boron IID processes were, therefore, estimated to be 6 dB cm^{-1} and 15 dB cm^{-1} respectively.

In view of the low losses attainable using fluorine indicated by the above measurements, longer (10 mm) fluorine disordered ridge waveguides were fabricated and the loss measured using the sequential cleaving technique.[3] This technique removes much of the uncertainty associated with end-fire coupling measurements. Total propagation losses as low as 4.7 dB cm^{-1} at a wavelength of 875 nm were obtained in these waveguides. A substantial (60 meV) blue shift accompanies a fluorine impurity induced disordering process with the process parameters and annealing conditions used here (dose of 10^{18} cm^{-3} annealed at 890 °C for 2 h). The total loss figure of 4.7 dB cm^{-1} is not the ultimate lower limit, since scattering due to rib waveguide top and sidewall roughness is likely to give an important contribution to propagation losses. This figure is, however, much lower than the contribution from free-carrier absorption ($\approx$ 40 dB cm^{-1}) expected from fully activated silicon doping at the concentrations typically required for QW disordering[22] ($> 10^{18}$ cm^{-3}).

Samples of the DQW epitaxial structure were also processed into waveguides with rib heights of nominally 0.5 μm and widths of 4 μm, both before and after disordering followed by annealing at 907 °C for 3 h. Sequential cleaving measurements on a non-implanted rib waveguide with this structure gave an estimated propagation loss of only 2.2 dB cm^{-1}, whereas that for a rib waveguide fabricated after implantation was 10.3 dB cm^{-1}. Measurements of the photoluminescence spectrum at 21 K for samples from the same wafer and disordered with the same process parameters show a blue shift of 90 meV in the excitonic peak of the QW. While the 10.3 dB cm^{-1} loss figure obtained is somewhat higher than the lowest figure obtained with the MQW structure, it is certainly low enough for applications such as passive waveguide sections, e.g. in integrated external cavity lasers[23] or DBR lasers. Furthermore, it is likely that a systematic exploration of the disorder processing conditions for such DQW structures would yield a significantly lower loss value, probably through using a somewhat lower annealing temperature than used here.

3.4. Near bandgap refractive index measurements

We have also carried out the first systematic studies of the effect of disordering on the refractive index.[6] The material structure investigated was again an MQW waveguide identical to that described above except that the MQW consisted of 54 periods of 60 Å GaAs wells and 60 Å $Al_{0.26}Ga_{0.74}As$ barriers. The absence of an upper cladding layer ensured a high optical field at the semiconductor-air interface for output coupling purposes. An electrochemical etch profile on the material showed the MQW region to be fully depleted at zero bias and the cladding layer to have a p-type background doping concentration of 1.8×10^{16} cm^{-3}.

Samples were again uniformly implanted throughout the depth of the MQW layer with boron or fluorine ions and were capped with a 1200 Å thick layer of plasma-deposited SiO_2 prior to annealing. In this case the thickness of this layer was designed to give a reasonable output coupling

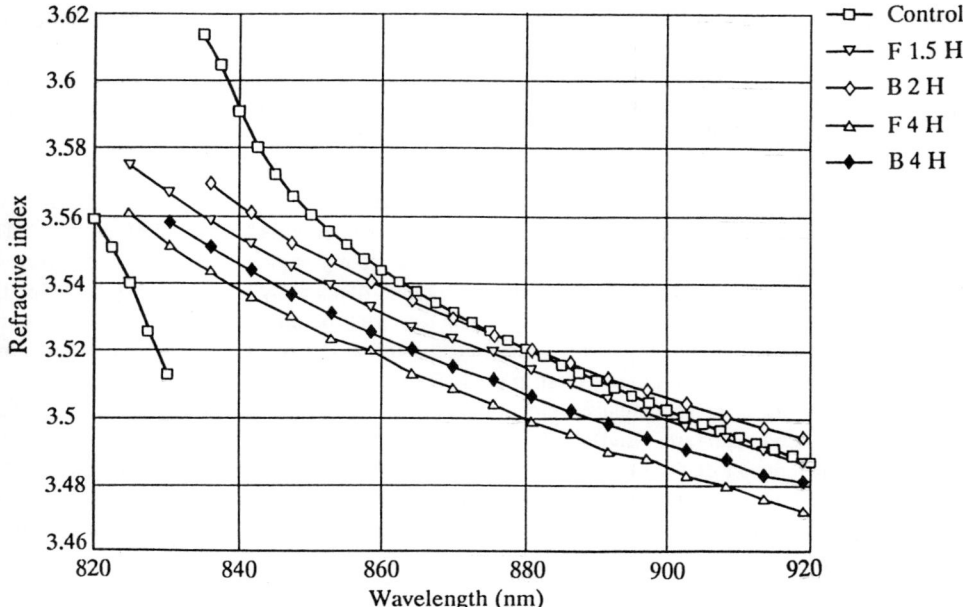

Fig. 5. Variation of refractive index with wavelength for the TE polarisation for boron disordered, fluorine disordered and control MQW (54 well) samples.

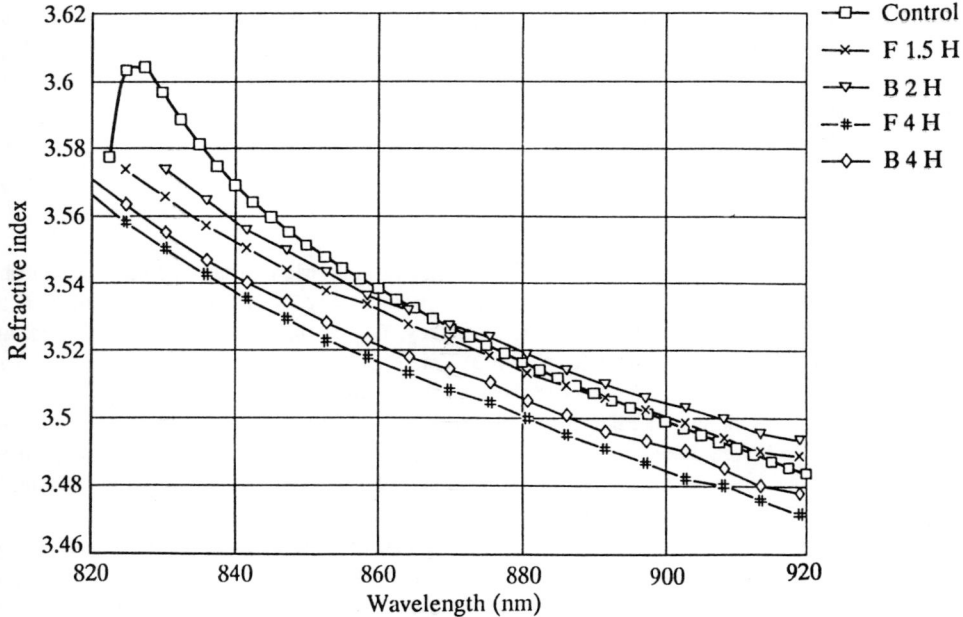

Fig. 6. Variation of refractive index with wavelength for the TM polarisation for boron disordered, fluorine disordered and control MQW (54 well) samples.

efficiency when used in the fabrication of an output coupler, as well as to give added protection against As desorption from the material. Annealing conditions used a temperature of 890 °C for times up to 4 h. Photoluminescence measurements showed an energy shift of 28 meV for boron after annealing for 120 min and of 40 meV for fluorine after 90 min. After 4 h with fluorine it is believed that the wells are completely disordered.

Output grating couplers were then fabricated in the SiO_2 annealing cap present on top of the slab waveguides. The grating pattern was produced by laser holography and transferred to the SiO_2 by shadow masking and dry-etching. The complete process is described in detail elsewhere.[24] The grating pitch of 285 nm was designed to give output coupling angles in the region of 10 to 45 degrees to the normal to the sample. Conventional optical lithography was then used to define the grating coupler region of the device as stripes 150 μm wide parallel to the grating and repeated periodically to give coupler devices of 3 mm length, with a 150 μm grating region and a 2 mm waveguide region before the grating to strip off leaky modes.

Waveguide index measurements were performed in the range 820 to 920 nm using a titanium:sapphire solid state laser. End-fire coupling was used to excite the waveguide mode(s) in the slab waveguide coupler device, and the output coupling angle from the grating was measured. The polarization dependence of the refractive index was investigated by varying the polarization of the Ti : Al_2O_3 laser beam. The modal refractive index of the slab waveguide was then found from the simple relation:[25]

$$n_g = \sin\Theta + m\frac{\lambda}{\Lambda} \qquad (3)$$

where n_g is the effective index of the waveguide, Θ is the output coupling angle, λ is the free space wavelength of the laser light, Λ is the pitch of the grating, and m is the order of the grating.

These results were used to find the material index of the MQW waveguide using a reiterative effective index procedure, and a Lorentzian oscillator model for 2-D excitons[26] provided a theoretical fit. The basic form of this model has previously been used for the modelling of MQW waveguide refractive indices:

$$\epsilon_{MQW}(\omega) = \epsilon_g(\omega) + 4\pi\beta_x \frac{\omega_x^2}{(\omega_x^2 - \omega^2 - i\omega\Gamma_x)} \qquad (4)$$

In the above equation, ϵ_g represents the background dielectric constant, containing contributions from all interactions except the exciton(s) in question. $\beta_{l,h}$ is the oscillator strength of the exciton transition, $\omega_{l,h}$ is the exciton centre frequency, $\Gamma_{l,h}$ is the linewidth of the exciton, and $\epsilon_{MQW}(\omega)$ is the dielectric constant for the MQW material. To model ϵ_g, a semi-empirical Sellmeier equation was employed to calculate the dielectric constant of $Al_xGa_{1-x}As$.[27]

The results are shown in Figs. 5 and 6, together with the refractive index results for the MQW waveguide before disordering. The range of results was limited by the tuning range of the Ti : Al_2O_3 laser in the case of the disordered material, whilst in the case of the control sample it is also limited by the absorption edge of the material at shorter wavelengths. The largest changes in the refractive index occur, as expected, at the exciton resonances in the starting material. At long wavelengths, the implanted samples annealed for short times are observed to have a higher refractive index than that of the starting material, this being particularly evident in the case of boron. After annealing for 4 h with fluorine the material refractive index is virtually identical for the two polarisations confirming that the MQW is completely disordered.

3.5. Diffusion of fluorine and boron

SIMS analysis of fluorine and boron disordered structures has also been carried out. Fig. 7 shows the variation of Ga and Al in the as-grown MQW structure and Fig. 8 shows the same structure after implantation with fluorine followed by annealing. The oscillations in the Al concentration are, as expected, washed out. (The longer period oscillations seen in both figures are an artefact of the sampling frequency of the SIMS system). Fig. 9 shows the effect of annealing on the fluorine distribution: rapid diffusion towards the surface and into the substrate takes place. In the case of boron, however, negligible diffusion takes place and the position of the three implants can still be seen even after annealing (Fig. 10).

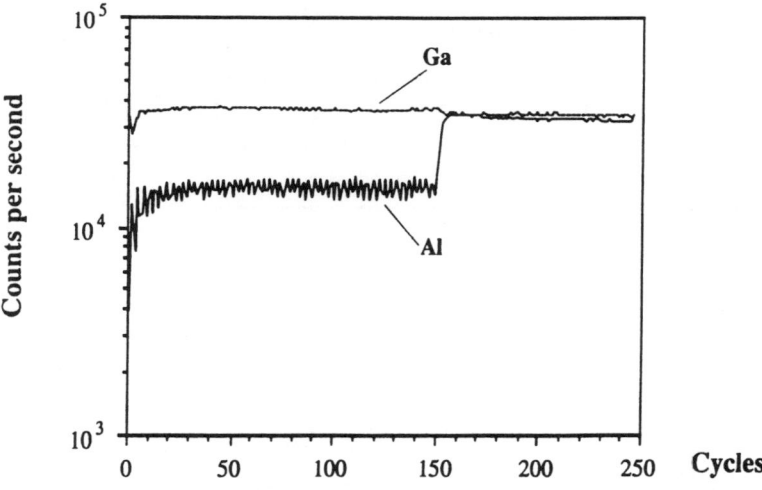

Fig. 7. SIMS analysis of the as-grown MQW structure.

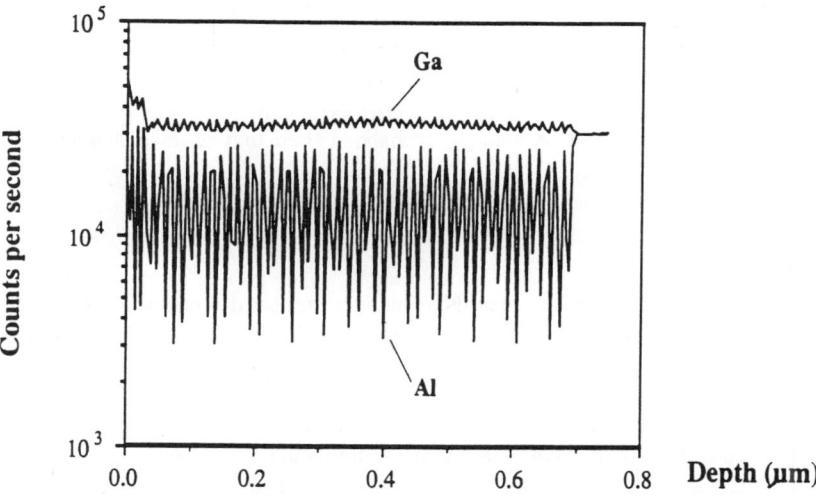

Fig. 8. SIMS analysis of a MQW sample after disordering with fluorine.

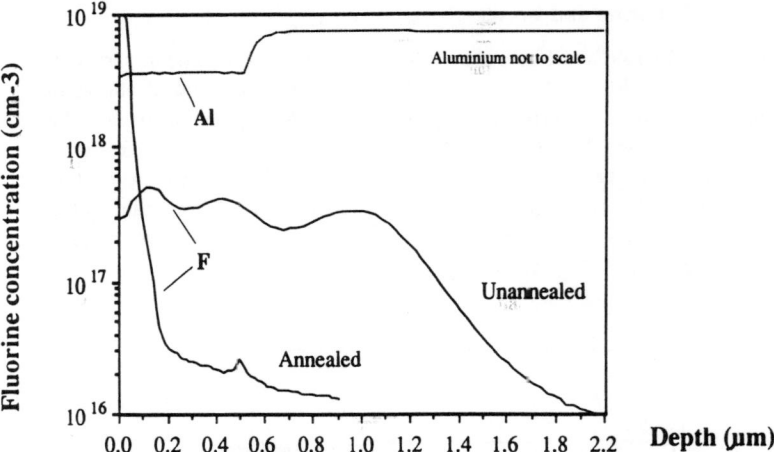

Fig. 9. SIMS analysis showing the diffusion of fluorine during an annealing cycle.

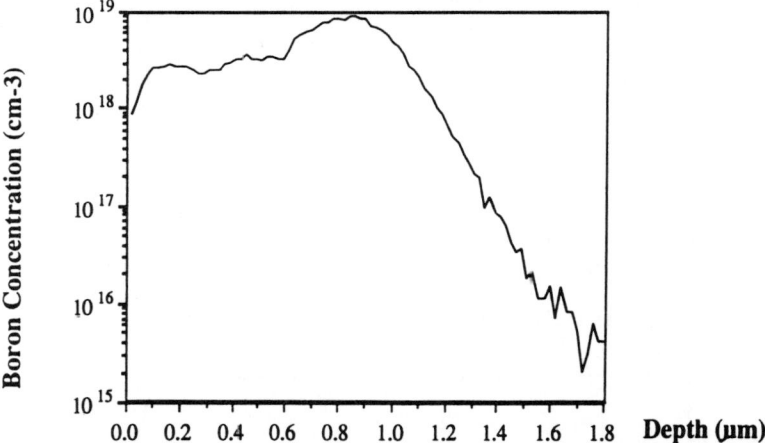

Fig. 10. SIMS analysis of a boron implanted sample after disordering. The distribution of the boron is virtually unchanged from that of as-implanted samples.

3.6. Electrical properties

The optical waveguide propagation loss measurements indicate that disordered material has a low free-carrier concentration. This was confirmed by electro-chemical profile measurements performed on the same 60 Å well / 60 Å barrier MQW structure as used in Sec. 3.4. An electrochemical profile indicated that the unimplanted MQW structure was fully depleted at zero applied voltage and that the residual doping of the AlGaAs cladding layer was 1.8×10^{16} cm^{-3} p-type. Samples were uniformly implanted to a depth of 1 μm with fluorine or boron to a concentration of 10^{18} cm^{-3} and annealed at 890 °C for 90 min and 120 min respectively. For these samples electrochemical profiles were as follows: the fluorine sample was p-type with a hole concentration of 1×10^{16} cm^{-3}, and the boron sample was also p-type with a hole concentration of 2.2×10^{16} cm^{-3}. These carrier densities would give rise to free-carrier absorption losses below 1 dB cm^{-1}, significantly lower than those observed in the waveguides. The resistivity of these layers is expected to be around 2 Ω·cm, a factor approximately 300 times lower than that of material disordered with silicon at a concentration of 10^{18} cm^{-3}. Good isolation resistances should therefore be realised within either fluorine or boron disordered waveguides with realistic practical dimensions.

4. IID of GaInAs/AlGaInAs and GaInAs/GaInAsP

4.1. Sample preparation

Disordering of two material systems for use at 1.5 μm has been investigated:[28,29] GaInAs/AlGaInAs and GaInAs/GaInAsP, both lattice-matched to InP. Three structures were investigated, two GaInAsP MQW structures—a separate confinement heterostructure (SCH) and a graded index structure (GRIN)—and an AlGaInAs MQW structure. The structures were grown by Atmospheric Pressure Metal-Organic Chemical Vapour Phase Epitaxy (APMOVPE) on n-type InP substrates. First a 1.0 μm n-type buffer layer of InP was grown then the QW structure. The P-quaternary materials contained four 60 Å $Ga_{0.47}In_{0.53}As$ wells separated by 120 Å $Ga_{0.17}In_{0.83}As_{0.37}P_{0.63}$ barriers with cladding layers on both sides of the wells. In the SCH structure the cladding layers were 0.09 μm thick $Ga_{0.17}In_{0.83}As_{0.37}P_{0.63}$ whilst in the GRIN the GaInAsP composition was graded from $Ga_{0.17}In_{0.83}As_{0.37}P_{0.63}$ at the MQW boundaries to InP at the outer edges. A further 1 μm layer of InP completed the P-quaternary structures. The Al-quaternary MQW consisted of four 100 Å $Ga_{0.47}In_{0.53}As$ wells separated by 50 Å $Al_{0.20}Ga_{0.27}In_{0.53}As$ barriers with a 200 Å layer of $Al_{0.20}Ga_{0.27}In_{0.53}As$ on top and a 1000 Å layer of the same alloy below. The exciton peak was measured by room temperature photoluminescence and found to be 1.522±0.016 for the SCH structure, 1.527±0.025 μm for the GRIN structure and 1.526±0.014 μm for the Al-quaternary material.

Samples of each structure were implanted with either fluorine or boron with a dose of 10^{14} cm^{-2}. Due to the non-uniformity of the thickness of the epilayers of the P-quaternary materials the implantation energy of the F$^+$ and B$^+$ had to be varied in order to stop the ions within the QW's: the implantation energy for F$^+$ was between 900 keV and 650 keV and for B$^+$ between 580 keV and 400 keV. In the case of the Al-quaternary, implants were made to two different depths, firstly to a depth of 30 nm (i.e. within the QW's) and secondly to a depth of 300 nm (i.e. ten times deeper than the wells). In order to implant ions into the QW's themselves it was necessary to cap the samples with a 100 nm layer of SiO$_2$ before implantation and the implant energies used in this case were 100 keV and 60 keV for the F$^+$ and B$^+$ respectively leading to concentrations of 5×10^{17} cm^{-3} in the wells. Whilst both implant species will knock silicon from the silica cap into the semiconductor, the calculated concentration of silicon atoms reaching the QW layer is about 10^{16} cm^{-3} for both fluorine and boron: a concentration too low to induce disordering.[28] The higher energy implants were made using F$^+$ at 250 keV and B$^+$ at 150 keV giving implant depths of 350 nm with a concentration in the wells of 2×10^{16} cm^{-3}. Calculations show that the damage in the wells was of similar magnitude for the two concentrations.

Three techniques were used for annealing of the P-quaternary: conventional furnace annealing in a high purity graphite box, furnace annealing with the samples in a partially evacuated silica ampoule and rapid thermal annealing (RTA). Conventional furnace annealing only was used for the

Al-quaternary. The degree of disordering was investigated by measuring the heavy hole exciton energy using PL at 15 K.

In the conventional annealing process samples capped with approximately 1000 Å of Si_3N_4 or SiO_2 were annealed for 0.5 to 2 h at temperatures ranging from 475 °C to 775 °C. A short set of experiments was run to investigate the effect of the annealing atmosphere on the disordering process. Samples were annealed for 0.5 h at 650 °C with a small quantity of indium phosphide loaded either with indium and tin, or with indium, tin and gallium, or indium phosphide and gallium arsenide loaded with gallium and tin in the graphite box, or with no other material in the box. The addition of the tin to the melt increases the solubility of the group V element (i.e. P or As), and so increases the partial pressure of the group V element within the graphite box to a level which prevents decomposition of the substrate or epitaxial layers.[30] It was found that there was no strong dependence either on the method used to create an overpressure or on the capping material.

To investigate further the effects of phosphorus overpressure on the exciton shift in the unimplanted material, uncapped samples were annealed in partially evacuated silica ampoules containing red phosphorus. The ampoules were annealed at 600 °C for 0.5 h. Samples annealed in the silica ampoules with no phosphorus and 6 mg of phosphorus showed shifts of 15 meV compared to 25 meV in the conventional graphite box anneal, samples annealed with 60 mg of phosphorus showed no shift.

4.2. Thermal Stability and Disordering of the 1.5 μm Materials Systems

Fig. 11 shows the results of annealing the P-quaternary SCH samples. It can be seen that the control samples suffer large exciton shifts at the temperatures of anneal investigated, which would lead to problems in fabricating integrated circuits by IID. The GRIN samples were annealed in the same way as the SCH samples and Fig. 12 shows the instability of the GRIN and SCH unimplanted samples at temperatures over 500 °C. It is believed that the observed blue shifts are caused by the diffusion of P into (and As out of) the wells.[31] For the Al-quaternary no shift in the exciton peak was observed for annealing temperatures up to 650 °C (Fig. 13), whilst above this temperature a red shift was obtained. This was probably caused by the Ga diffusing out of the wells and being replaced by In from the barriers. This is unexpected since the In concentration is initially constant but this effect has previously been observed in GaInAs/AlInAs QW structures.[32,33] The effect is probably caused by the low mobility of the Al compared to the Ga and In and the requirement of the material to remain stoichiometric. Since the aluminium does not diffuse significantly the In diffuses to maintain the structure of the material.

The boron implanted P-quaternary produced a red shift after annealing at temperatures above 600 °C (Fig. 11), which can be explained by diffusion occuring only on the group III sublattice. The In concentration in the wells would increase therefore decreasing the exciton energy. At lower annealing temperatures small blue shifts were measured probably caused by disordering due to the damage from the implantation process. Boron implanted Al-quaternary samples exhibited small increases in the exciton peak energy of a few meV for annealing temperatures below 650 °C (Fig. 14). Similar shifts were measured for the two different concentrations. This suggests that the intermixing caused by the boron implants is due to the damage incurred during implantation. At higher annealing temperatures blue shifts were observed in the wells with the higher concentration of boron suggesting that at these higher temperatures the boron may become active in the disordering process.

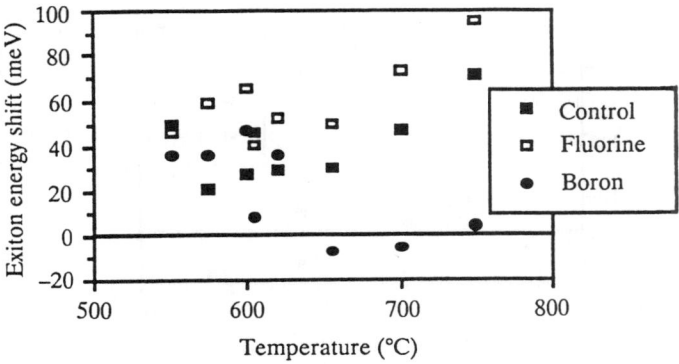

Fig. 11. The exciton peak shift after annealing for 30 min as a function of annealing temperature for the unimplanted and the boron and fluorine implanted P-quaternary SCH material.

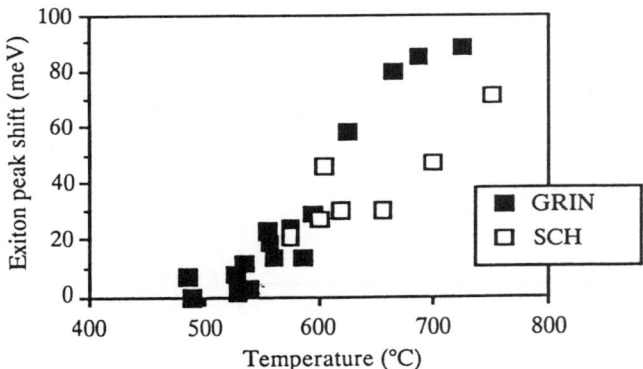

Fig. 12. The exciton peak shift after annealing for 30 min as a function of annealing temperature for the unimplanted P-quaternary SCH and GRIN material.

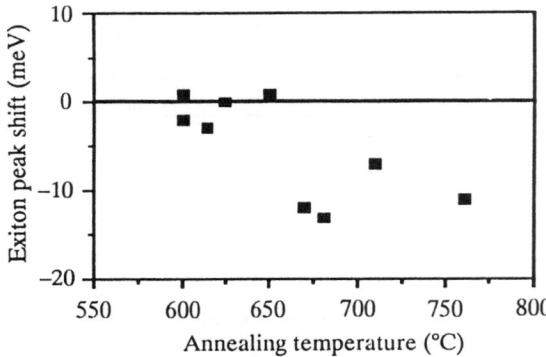

Fig. 13. The exciton peak shift after annealing for 30 min as a function of annealing temperature for the unimplanted Al-quaternary material.

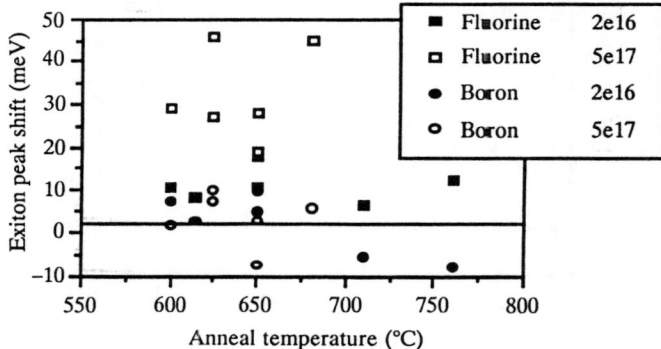

Fig. 14. The exciton peak shift after annealing for 30 min as a function of annealing temperature for the boron and fluorine implanted Al-quaternary material.

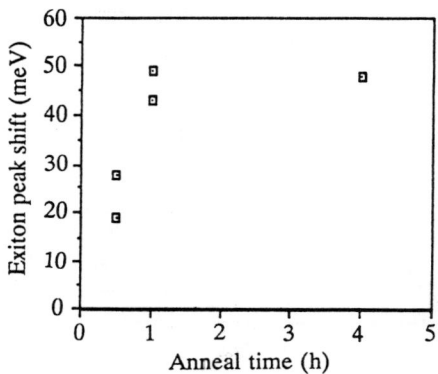

Fig. 15. The exciton peak shift as a function of annealing time at 650°C for fluorine implanted Al-quaternary material.

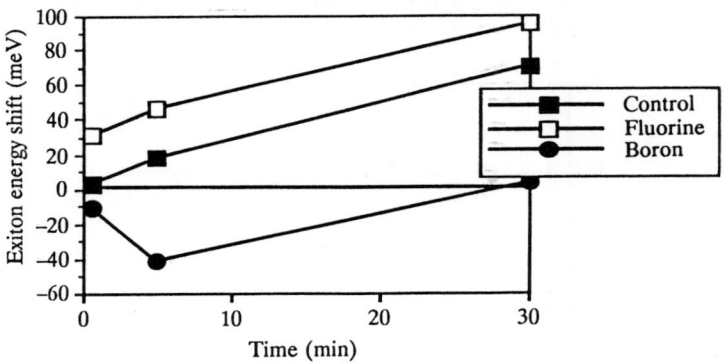

Fig. 16. The exciton peak shift as a function of annealing time at 750°C for unimplanted, fluorine implanted and boron implanted P-quaternary material.

From Fig. 11 it can be seen that at low annealing temperatures the fluorine caused a blue shift about twice the size of that caused by boron in the P-quaternary. Calculations show that fluorine induces approximately twice as much damage as boron consistent with observing a larger shift. For annealing temperatures above 650 °C, the fluorine implanted samples produce a shift approximately 20 meV larger than the control samples, suggesting thermal effects are controlling the disordering rather than impurity induced effects. From Fig. 14 it can be seen that blue shifts of up to 45 meV were obtained in the Al-quaternary using fluorine. The shifts in the exciton peak for the samples implanted with the higher concentration of fluorine were considerably larger than those observed in the boron implanted material. This suggests that fluorine is active in disordering the wells and produces larger shifts than damage induced effects. Further evidence for this was that the lower concentration of fluorine produced blue shifts at annealing temperatures above 650 °C where the unimplanted material exhibited red shifts.

By annealing the fluorine implanted Al-quaternary for longer times at 650 °C the shift is increased to about 45 meV where it saturates (see Fig. 15). The average composition of the well region was calculated to be $In_{0.52}Ga_{0.40}Al_{0.08}As$. Published photoluminescence[34,35] data at 10 K indicate that the exciton peak of this alloy is 30 meV larger than that of the exciton peak in the wells of the undisordered material. This is the expected maximum shift in the exciton peak if the wells and the barriers were totally intermixed. The magnitude of the observed shifts therefore suggest that aluminium from the cladding layer may also be diffusing into the well region, increasing the aluminium content of the resulting alloy and leading to a larger than expected blue shift in the exciton energy.

Samples of the P-quaternary with and without implants were annealed for 0.5, and 5 min at 750 °C in an RTA system. The results, illustrated in Fig. 16, show that the control sample gave a shift less than +5 meV for a 0.5 min anneal, and 18 meV after 5 min. The boron implanted samples gave a red shift of 40 meV after a 5 min anneal, while fluorine implanted samples always produced a blue shift. A shift of 30 meV was obtained by annealing at 750 °C for 0.5 min. The exciton peaks were clear indicating that most of the implantation damage had been annealed out, however there was some evidence of the creation of deep levels in the cladding layers.

4.3. Comparison of Al- and P-Quaternaries

The P-quaternary disorders without any implants at annealing temperatures above 500 °C. It is thought that P diffuses into and As out of the wells producing a blue shift in the exciton peak. The Al-quaternary is stable up to annealing temperatures of 650 °C. Above this temperature the exciton peak red shifts probably due to the interdiffusion of Ga and In. The better temperature stability of the Al-quaternary makes it a more attractive material for IID processing.

At lower annealing temperatures (600 °C for the P- and 650 °C for the Al-quaternary) boron causes some intermixing probably due to the damage caused by implantation. At higher temperatures there is evidence that it plays an active role in disordering the group III sublattice, producing a red shift in the P-quaternary and a blue shift in the Al-quaternary. The damage caused by implanting fluorine into the P-quaternary appears to be responsible for the intermixing at low annealing temperatures. Above 600 °C it produces a larger blue shift than found with the control samples but still appears to be dominated by thermal disordering processes. Fluorine produces significant blue shifts in the exciton peak of the Al-quaternary at all annealing temperatures investigated suggesting it has an active role in the disordering process.

Rapid thermal annealing looks to be a promising method for processing the P-quaternary. It was possible to produce a significant blue shift in material implanted with fluorine while, under the same annealing conditions, the control samples remained unchanged.

5. Conclusions

Impurity induced disordering using neutral impurities in the GaAs/AlGaAs system has been demonstrated to be a versatile technology for use in the fabrication of the high performance components and integrated devices needed for coherent optical systems. All of the loss criteria identified in the applications outlined in Sec. 2 can be met in the GaAs/AlGaAs system: by using fluorine disordering, propagation losses as low as 4.6 dB cm^{-1} have been demonstrated. The refractive index changes associated with IID have been systematically measured for the first time.

The work described in this paper is mainly concerned with the GaAs/AlGaAs system; most systems applications however require device operation at wavelengths around 1.5 μm. Studies have now been made of the bandgap changes associated with boron and fluorine IID in both the GaInAs/AlGaInAs and GaInAs/GaInAsP systems lattice matched to InP. The Al-quaternary appears to be more temperature stable than the P-quaternary and would be preferred for IID processing. In both systems fluorine (but not boron) is an active disordering species resulting in bandgap increases. The potential of IID in these material systems is considerable.

Acknowledgments

The support of Prof R.M. De La Rue and is gratefully acknowledged. Samples were grown by C. Button and J.S. Roberts of the University of Sheffield and by R.W. Glew of Bell Northern Research Europe. This work was supported by SERC under several grants.

References

[1] J.H. Marsh 1988 *SPIE Proc* **861** 118
[2] D.G. Deppe and N. Holonyak Jr 1988 *J. Appl. Phys.* **64** R93
[3] M. O'Neill, J.H. Marsh, R.M. De La Rue, J.S. Roberts, and R. Gwilliam 1990 *Electron. Lett.* **26** 1613
[4] M. O'Neill, A.C. BRYCE, J.H. Marsh, R.M. De La Rue, J.S. Roberts, and C. Jeynes 1989 *Appl. Phys. Lett.* **55** 1373
[5] J.P. Van Der Ziel, M. Ilegems, and R.M. Mikulyak 1976 *Appl. Phys. Lett.* **67** 735
[6] S.I. Hansen, J.H. Marsh, J.S. Roberts, and R. Gwilliam 1991 *Appl. Pays. Lett.* **58** 1398
[7] I.J. Pape, P. Li Kam Wa, J.P.R. David, P.A. Claxton, and P.N. Robson 1988 *Electron. Lett.* **24** 1217
[8] I.J. Pape, P. Li Kam Wa, D.A. Roberts, J.P.R. David, P.A. Claxton, and P.N. Robson 1988 *GaAs and Related Compounds. (Inst. Phys. Conf.)* Ser No 96 397
[9] M. Razehgi, O. Archer, and F. Launay 1987 *Semicond. Sci. Technol.* **2** 793
[10] K. Nakashima, Y. Kawaguchi, Y. Kawamura, and Y. Imamura 1988 *Appl. Phys. Lett.* **52** 1383
[11] I.J. Pape, P. Li Kam Wa, J.P.R. David, P.A. Claxton, P.N. Robson, and D. Sykes 1988 *Electron. Lett.* **24** 910
[12] M.A. Bradley, F.H. Julien, J.P. Gilles, Y. Gao, E.V.K. Rao, M. Razeghi, and F. Omnes 1990 *Electron. Lett.* **26** 209
[13] B. Tell, B.C. Johnson, J.L. Zyzkind, J.M. Brown, J.W. Sulhoff, K.F. Brown-Goebler, B.I. Miller, and U. Koren 1988 *Appl. Phys. Lett.* **52** 1428
[14] H. Sumida, H. Asahi, S. Jae Yu, K. Asami, S. Gonda, and H. Tanoue 1989 *Appl. Phys. Lett.* **54** 520
[15] B. Tell, J. Shah, P.M. Thomas, K.F. Brown-Goeblere, A.D. Giovanni, B.I. Miller, and U. Koren 1989 *Appl. PhysLett.* **54** 1570
[16] M. Suzuki, H. Tanaka, S. Akiba, Y. Kushiro 1988 *J. Lightwave Technol.* **6** 779
[17] P.W.A. MC Ilroy, A. Kurobe, and Y. Uematsu 1985 *IEEE J. Quantum Electron.* **QE-21** 1958
[18] G.H.B. Thompson 1980 *Physics of Semiconductor Laser Devices* (Wiley)
[19] J.D. Ralston, L.H. Camnitz, G.W. Wicks, and L.F. Eastman 1986 *GaAs and Related Compounds (Inst Phys Conf)* Ser No 83 367
[20] J.S. Roberts, M.A. Pate, P. Mistry, J.P.R. David, R.P. Franks, M. Whitehead, and G. Parry 1988 *J. Cryst. Growth* **93** 877

[21] J.D. Ralston, S. O'Brien, G.W. Wicks, and L.F. Eastman 1988 *Appl. Phys. Lett.* **52** 1511
[22] R.L. Thornton, W.J. Mosby, and T.L. Paoli 1987 *IEEE J.Lightwave Technol.* **LT-6** 786
[23] J. Werner, E. Kapon, N.G. Stoffel, E. Colas, S.A. Schwarz, C.L. Schwarz, and N. Andreadakis 1989 *Appl. Phys. Lett.* **55** 540
[24] S.I. Hansen, J.H. Marsh, and J.S. Roberts *IEE Proceedings Part J* (to be published)
[25] M.L. Dakss, L. Kuhn, P.F. Heidrich, and B.A. Scott 1970 *Appl. Phys. Lett.* **16** 525
[26] M. Dagenais and W.F. Sharfin 1985 *J. Opt. Soc. Am.* **B2** 1179
[27] M.A. Afromowitz 1974 *Solid State Commun.* **15** 59
[28] A.C. Bryce, J.H. Marsh, R. Gwilliam, and R.W. Glew 1991 *IEE Proc. Part J (Optoelectronics)* **138** 87
[29] J.H. Marsh, S.A. Bradshaw, A.C. Bryce, R. Gwilliam, and R.W. Glew *submitted to J. Electron. Mat.*
[30] G.A. Antypas 1980 *Appl. Phys. Lett.* **37** 64
[31] K. Nakashima, Y. Kawaguchi, Y. Kawamura, H. Ashasi, and Y. Imamura 1987 *Jpn. J. Appl. Phys.* **26** L1620
[32] R.J. Baird, T.J. Potter, G.P. Kothiyal, and P.K. Battacharya 1988 *Appl. Phys. Lett.* **52** 2055
[33] R.J. Baird, T.J. Potter, R. Lai, G.P. Kothiyal, and P.K. Battacharya 1988 *Appl. Phys. Lett.* **53** 2302
[34] J.P. Praseuth, M.C. Joncour, J.M. Gerard, P. Henoc, and M. Quillec 1988 *J. Appl. Phys.* **63** 400
[35] J.I. Davies, A.C. Marshall, M.D. Scott, and R.J.M. Griffiths 1988 *Appl. Phys. Lett.* **53** 276

Reactive ion etching for fabrication of intergrated optic and optoelectronic elements

F.N. Timofeev

A.F. Ioffe Physico-Technical Institute, Academy of Sciences of the USSR,
26 Polytekhnicheskaya st. 194021 Leningrad, USSR

Abstract. A reactive ion etching process (RIE) of GaAs(AlGaAs), InP with Cl_2; $Cl_2:CH_4 + Ar$; CCl_2F_2; BCl_3 mixtures is described. The influence of the process parameters on the etching rate is given for GaAs, GaAlAs and InP. The variations of the RIE process parameters take possibility to change the side wall geometry up to the vertical. The damage caused by RIE was studied by Raman spectroscopy and PL. The results of application $As_2S_3(As_2S_3:Zn)$/photoresist and other multilayer masks for submicron-size element formation are presented.

1. Introduction

The major problem in Reactive Ion Etching (RIE) of III–V semiconductors has been, and remains, the precision defect-free reproducible etching of semiconductor materials with predictable etching profiles and selectivity. In this paper only some details of AlGaAs/GaAs, InP structure RIE micromachining are discucced. The first problem is the formation of a precision sub-microne size mask and application of the relatively well known As_2S_3 material and of a quite new material, such as $As_2S_3:Zn$ for this purpose. The use of different gases (CCl_2F_2; BCl_3; Cl_2; Cl_2+CH_4) for precision isotropic (cristallografic) and anisotropic RIE of GaAlAs/GaAs structures are the second topic of this work. Finally, some data on surface damage characterization (due to the RIE of GaAs) will be presented. The data were obtained by Raman spectroscopy, photoluminescense (PL) and the PL of multiple-quantum-well (MQW) structures.

2. RIE processing

The RIE was performed in a planar diode-type "Alcatel RDE-300" load-locked machine, operated at 13.56 MHz RF. The base pressure of this reactor was about 10^{-6} Torr. In diode-type RIE the ionisation rates were usually very low ($< 10^{-4} \div 10^{-5}$). So the gas mixture in the RIE reactor consisted of a lot of neutrals and some portion of ions and radicals. Positive ions and radicals were accelerated in Self-Bias (SB) electric field and directed to the substrate (etched sample). The energy of the ions and radicals ranged from a few electronvolts up to several hundreds of electronvolts and depended on SB voltage, pressure in the chamber, type of gases, etc. In generall, optically active elements require anisotropic etching, while selectivity is more important in transistor fabrication. The anisotropy is refered as to preferential erosion in a direction being normal to the surface of a wafer. Anisotropic etching can be achieved by different ways. The first way is cryogenic RIE.[1] In this case the temperature of substrate holder is decreased down to $-40 \div -100\,°C$, so thermally activated neutral reactions are suppressed and only energy-driven, ion-assisted etching

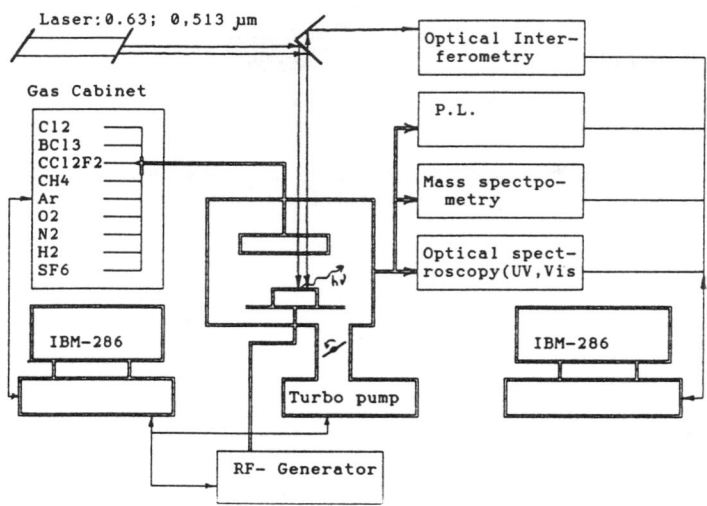

Fig. 1. Schematic diagram of the RIE system.

reaction occurs. The second way is inhibitor-driven ion-assisted etching.[2] In this case a protective sidewall film prevents the etching in the horizontal direction, and only vertical, quasi-anisotropic etching takes place. The protective sidewall film may originate from involatile etching products or from film-forming precursors which are adsorbed during the etching process.[2] Anisotropic etching can be also achieved by pressure reduction and SB voltage increase.

The analytical techniques (Fig. 1) attached to our RIE reactor included mass spectrometer (Balzers model QMS 112A) for chamber contamination control, visible and UV plasma spectroscopy (Sofie Instruments Multisem 440), laser interferometry of 0.63 μm and 1.06 μm wavelength (Sofie Instruments Multisem 440) for etching rate measurements and home-made in-chamber optical spectrometry for *in situ* control of the AlGaAs layers temperature.[3] The RIE machine operation was achieved using a home-made process control system based on IBM PC/AT. This system provided automatic start and operation of the RIE machine and showed excellent results in low SB voltage (5 ÷ 50 V) etching experiments. In our experiments we investigated the etching rate of different materials as a function of plasma SB voltage (SB measurements provided more reproducible results of etching as compared with the case when RF power was kept constant).

Typically 5 ÷ 30 mm square samples were used for etching. The samples were put onto the water-cooled cathode without heat conductive paste. All the etching rates reported here were obtained by interferometry or using a Dektak stylus profilometer. Scanning electron microscopy was used to examine the surface morphology and etching profiles after the RIE. After the etching the samples were kept in a dry N_2 glove box until they could be transfered to the various measurement systems for analysis.

3. Single and multilayer mask for RIE

3.1. Single layer mask

This kind of masks is the simplest in preparation and consist of a single AZ-type photoresist layer

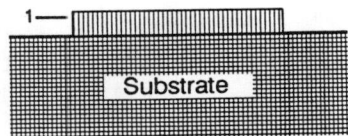

Fig. 2. Schematic illustration of the single layer mask. 1—radiation sensitive RIE resistant photoresist mask.

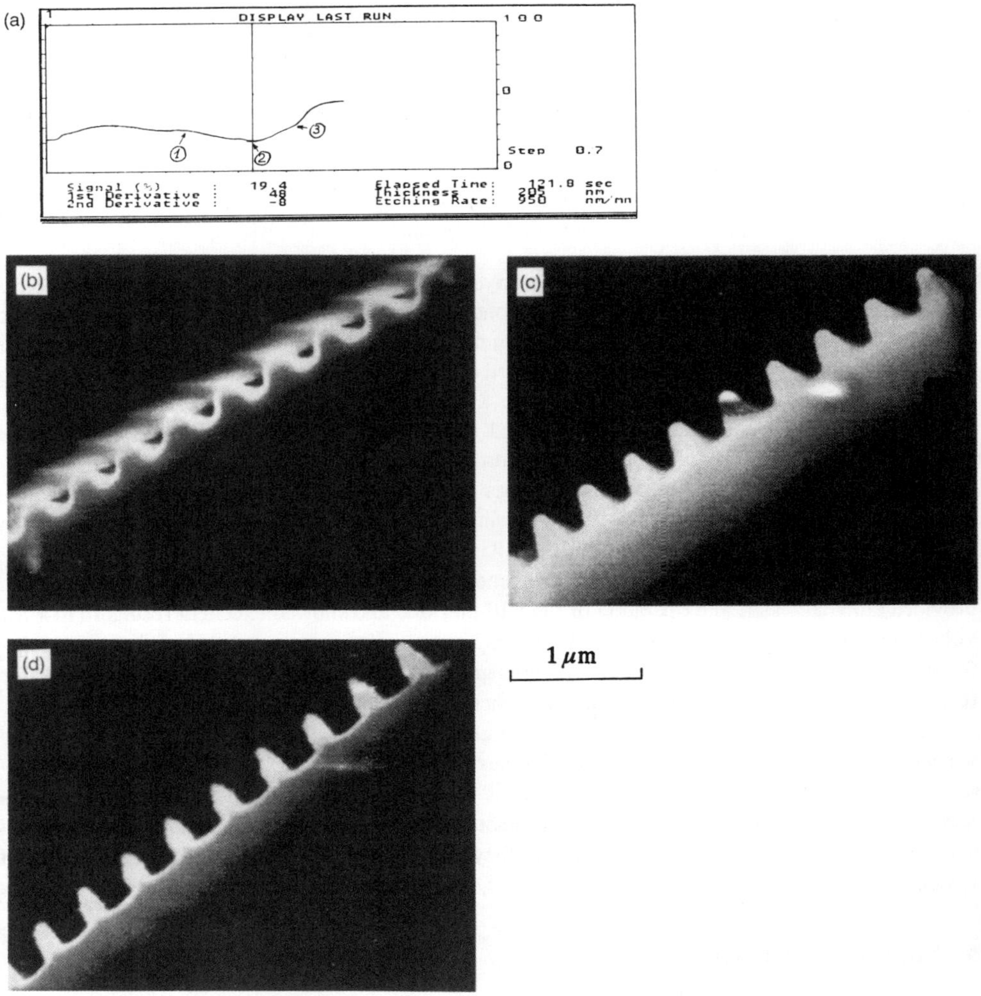

Fig. 3. (*a*) interferogramm of GaAs DG formation by RIE process. SEM photo of DG etched for 90 s (*b*), 121 s (*c*); and 154 s (*d*) in 4 sccm Cl$_2$ plasma; U$_{SB}$ = −100 V; 0.5 Pa in-chamber pressure.

(Fig. 2). The AZ-type photoresist, usually diluted with AZ thinner was spun at the sample surface to obtain a mask 100–2000 nm in thickness. After a standard photolithographic procedure and development, the photoresist mask was post-baked at 90 °C for 30 min. As a result, the mask was usually the thinnest at the edge of an opening (the edge of mask angled), so after the RIE the actual dimensions of a semiconductor device were changed.[2] Such single layer mask can be successfully applied for the formation of stripes and plane diffraction gratings (DG). Usually a grating is formed by chemical or reactive ion etching through a photoresist mask realized by holographic technique or electron beam exposure. The thickness of the mask depends on its period and varies over the range $300 \div 100$ nm (for AlGaAs DFB/DBR semiconductor lasers with the 1-st and the 2-nd order distributed feedback). In case of holographic exposure the thickness of the resist was determined to reduce the influence of standing waves of exposure light in the resist layer. Fig. 3 shows the results of the RIE of AlGaAs through DG AZ 1450-type mask of 0.47 μm period. In order to prevent overetching of a DG, we controlled the photoresist mask thickness and DG reflectivity by laser interferometry. In Fig. 3(a) an interferogramm of the RIE process is shown. The point 1 on this plot corresponds to the shallow DG (see SEM photograph of the grating in Fig. 3(b)). The point 2 on interferogramm corresponds to the DG with the highest diffraction efficiency, or the maximal grating depth (the SEM photograph of the grating is shown in Fig. 3(c)). At this point the DG reflectivity is minimal, because of a large part of the incident light reflected in diffraction orders. Further etching of the sample results in overetching and transformations of DG profile to "finger-type" shape (Fig. 3(d)). The diffraction efficiency at the point 3 (Fig. 3(a)) is decreased and reflectivity of DG surface is increased.

3.2. Multi-layer mask

Three-level systems are more flexible than a single-layer mask. Using such systems one can overcome problems associated with the wafer topography (step coverage) and surface reflectivity (resolution).[4] A schematic representation of the three-level mask processing is shown in Fig. 4(a). After pattern definition of an upper radiation-sensitive layer, the pattern is transferred to an optional intermediate layer (usually a metal layer) by a wet or dry etching process. Finally, the pattern definition of oxygen-RIE resistant intermediate layer is transferred to the substrate by the oxygen-RIE technique. At this stage of the mask processing we used He–Ne laser interferometry to control the "stop-etch" point of thick planarizing RIE-resistant bottom resist layer. Fig. 5(a) shows a SEM micrograph of the completed three-level mask with 80 nm lines and 1000 nm spaces. In this case the initial pattern was performed by holographic exposure. Loss in linewith was less than 15 nm. A AZ-1375 photoresist was used as a planarizing layer. In some cases polymide films can be used as planarizing layer in a three-level mask. Fig. 6 shows a comparison of the etching rate of a hard-baked AZ-1375 photoresist with that of a polymide film as a function of SB voltage in oxygen plasma. Polymide films have a relatively high etching rate compared to AZ-1375 films. Fig. 5(b) showns a SEM micrograph of a three-level mask with a polymide planarizing layer. The initial pattern was performed by UV photolithography.

Three-level masks have the potentiality for improved resolution inherent in the anisotropic etching. However, three-level process has a significant number of processing steps and requires a precision control of the spin-rotation coating step to prevent large scale nonuniformities of the photoresist layers. Incorporating the properties of the top imaging layer with those of the oxygen-RIE resistant, intermediate inorganic layer into a single imaging layer would simplify this system.

Two-level mask (Fig. 4(b)) consists of a single, radiation sensitive RIE resistant layer, coated onto the surface of thick planarizing layer. A conventional processing allows pattern def-

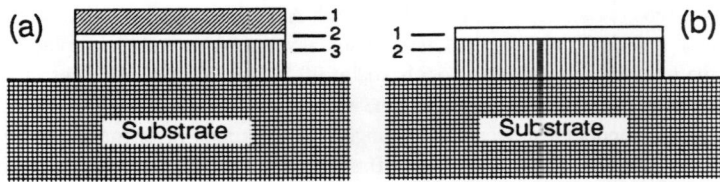

Fig. 4. (a) schematic illustration of the three-level mask for RIE, 1—the thin radiation sensitive layer; 2—the oxygen-RIE resistant layer; 3—thick planarizing RIE-resistant layer; (b) schematic illustration of the two-level mask for RIE. 1—radiation sensitive oxygen-RIE resistant layer; 2—thick planarizing RIE-resistant layer.

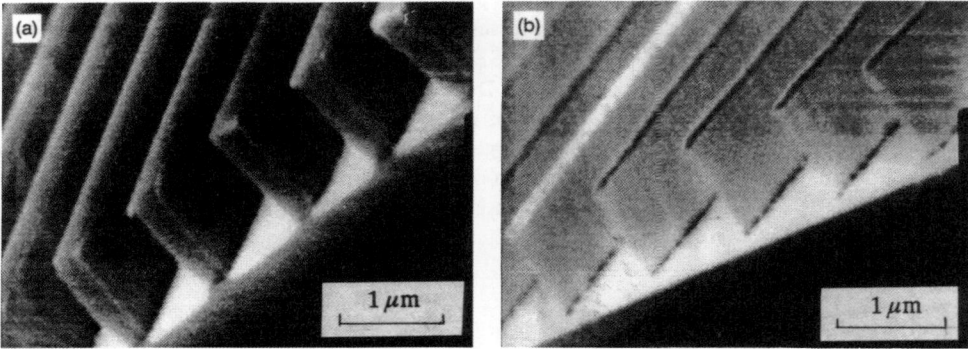

Fig. 5. SEM micrograph of completed tree-level masks with a AZ-1375 planarizing layer (a) and a polymide film planarizing layer (b).

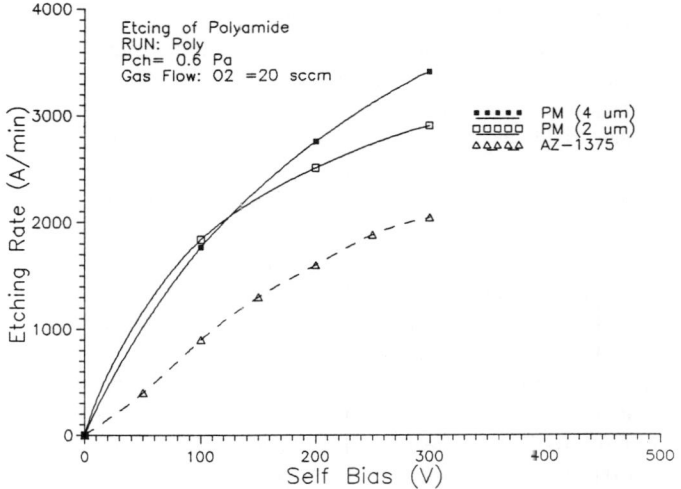

Fig. 6. Comparison of the etching rate data of the AZ-1375 photoresist with that of a polymide film as a function of SB voltage in O_2 plasma.

inition of this upper layer, and the pattern is then transferred to the substrate by the oxygen-RIE technique. In our experiments we used the silicon-containing photoresist as a top layer. Parameters of this two-level mask and the processing are similar to those described by E. Reichmanis et al.[4] In this case both layers were made by spin rotation coating. The resulting double-layer film usually was varied in thickness. The variations are especially large at the edge of the substrate, so after the mask formation they can affect the uniformity of the pattern. The double-layer mask with oxygen-RIE resistant top sensitive layer coated by sputtering are more attractive. We have demonstrated the application of two-level As_2S_3/photoresist mask for submicrone RIE.

4. Application of As_2S_3/photoresist two-level mask for submicrone RIE

It is known that amorphous chalcogenides are inorganic resists with high resolution.[5] The resist effect in these materials is due to photostructural changes under illumination, for example under light radiation. The main advantages of these materials are their good sensitivity to different types of radiation (visible or UV light, electron and ion beams, x rays) and the possibility to prepare thin uniform films on nonplanar substrates by vacuum thermal evaporation.

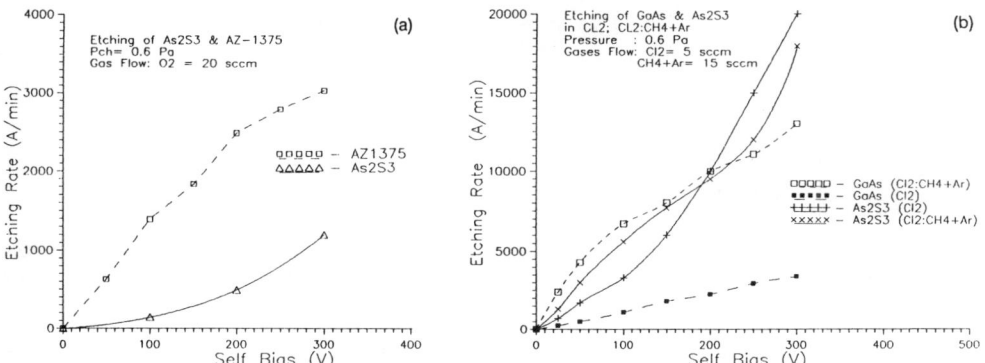

Fig. 7. Comparison of the etching rate of an As_2S_3 film with the hard baked AZ-1375 photoresist as a function of SB voltage in O_2 plasma (a); and of an As_2S_3 film with GaAs as a function of SB voltage in Cl_2 and $Cl_2:CH_4 + Ar$ plasmas (b).

In our experiments As_2S_3 films (0.2 ÷ 5.0 μm) were made in a thermal sputtering machine VUP-5. The thickness of As_2S_3 films was measured in the chamber by laser interferometer or quartz thickness monitor. The temperature of the evaporator was 420 °C. Oxygen RIE was performed at low pressure of 0.6 Pa to provide high anisotropic etching regime. Fig. 7(a) shows a comparison of the etching rate data of As_2S_3 film with that of the hard-baked AZ-1375 photoresist as functions of SB voltage in oxygen plasma. The etching rate of As_2S_3 is 3 ÷ 7 times lower than that of AZ-1375. The etching rates increase with SB because of the greater sputter component of the etch at higher bias voltage. The etching at SB voltage > -150 V is more anisotropic. Fig. 7(b) shows a comparison of etching rates of As_2S_3 film with those of GaAs as functions of SB voltage in Cl_2 and Cl_2+CH_4:Ar plasma. The RIE was performed at low pressure of 0.5 ÷ 0.6 Pa and gas flow: $Cl_2 = 5$ sccm; CH_4:Ar(1:5)= 15 sccm. The etching rates of As_2S_3 at SB > -100 V are practically the same as for GaAs. However, the etching rates of the AZ-1375 photoresist under the same condition are significantly lower. So in double-layer As_2S_3/AZ-1375 mask the thin top

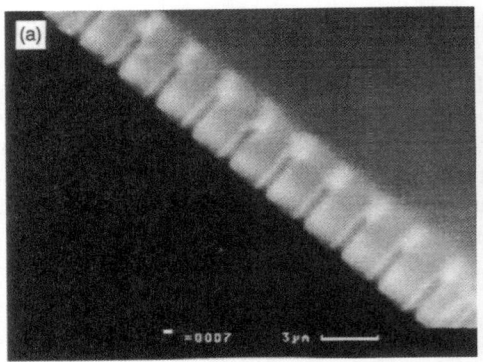
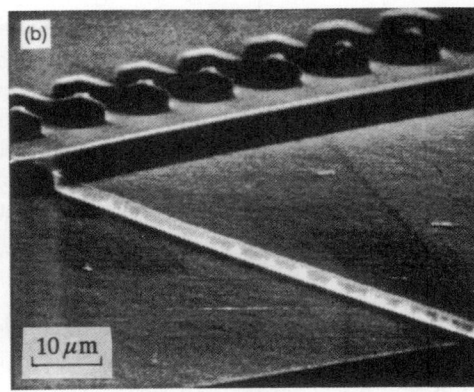
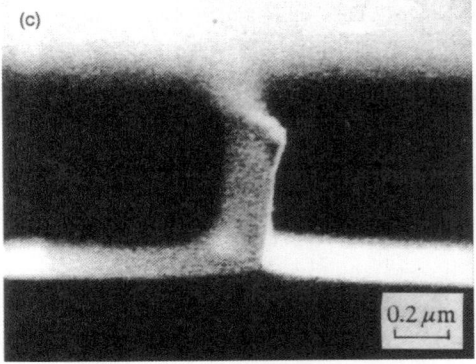

Fig. 8. SEM photographs of GaAs structures etched in $Cl_2:CH_4+Ar$ plasma through the As_2S_3/AZ-1375 two-layer mask. The masks were patterned by UV photolithography (a) and e-beam lithography (b,c).

As_2S_3 layer is removed in Cl_2 plasma during a few second, or in other words, this As_2S_3 layer is self-removable. Fig. 8(a) shows a SEM picture of a periodic structure, made in GaAs by the Cl_2+CH_4:Ar RIE through the As_2S_3/AZ-1375 mask. The pattern in As_2S_3 layer was made by UV photolithography. Then the As_2S_3 was developed in an organic developer. Finally, the AZ-1375 bottom layer was etched through the As_2S_3 mask by O_2-RIE. The result of GaAs RIE through the same mask is shown in Fig. 8(b). In this case the pattern was made by e-beam lithography. The elements less then 60 nm in size were made with the dimentional accuracy of ± 10 nm (Fig. 8(c)). The minimal size of the elements was limited by the resolution of our e-beam lithography.

A chalcogenide-based mask enables high uniformity[6] and reproducibility due to high degree of planarity of the chalcogenide layer on the non-planar organic resist layer (AZ-1375).

As_2S_3:Zn mask. Zn–doped As_2S_3 quasi-single-layer mask shows a very interesting feature. This mask consists of a thin As_2S_3 layer made by thermal evaporation ($0.1 \div 0.4\,\mu m$). In this radiation sensitive layer the pattern was made by e-beam lithography. After the etching of As_2S_3 in organic developer, thin layer of Zn (< 100 nm) was sputtered onto the patterned substrate. Under thermal annealing at 120 °C during 30 min Zn had diffused inside the As_2S_3 film. Fig. 9 shows a SIMS profiling of this structure. Zn diffused in As_2S_3 quite uniformly. The presence of Zn impurity in the As_2S_3 film strongly changes the etching rate in Cl_2 plasma (Fig. 10(a)). The etching rate of As_2S_3:Zn in Cl_2 plasma decreases; at the same time the etching rate of this material in O_2 plasma does not change as compared to pure As_2S_3. Fig. 10(b) shows a SEM picture of GaAs etched through the As_2S_3:Zn mask. We believe, that this kind of masks can be applied for nanometer-scale element's formation. Since all the technological steps of the mask formation are of planar type (thermal evaporation, diffusion), excellent thickness uniformity of the mask has been achived. This provides to improve the pattern element uniformity.

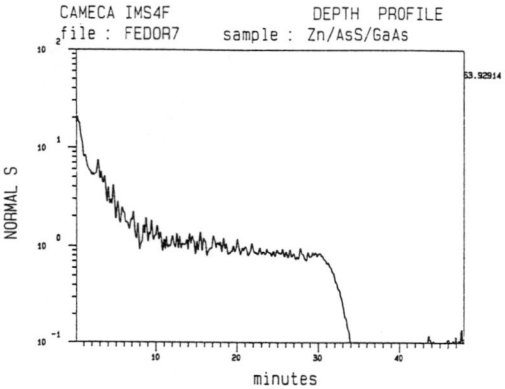

Fig. 9. Depth distribution of Zn in 0.4 μm thick As_2S_3 film obtained by SIMS profiling (33 min of etching corresponds to 0.4 μm).

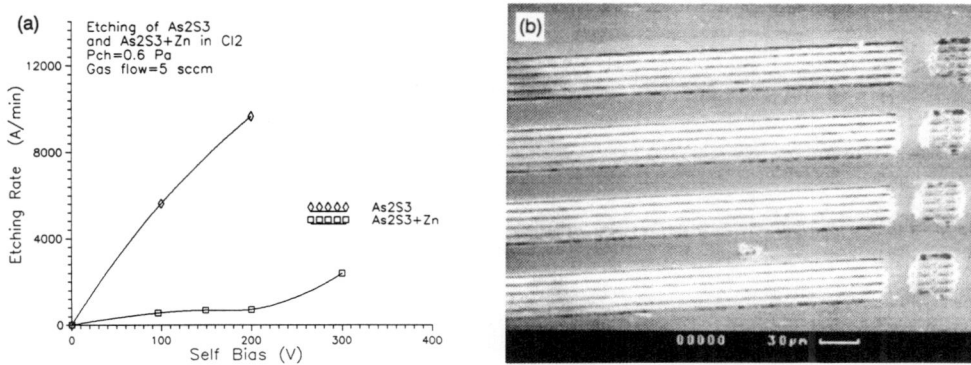

Fig. 10. (a) comparison of the etching rate data on As_2S_3 with an As_2S_3:Zn film as functions of SB voltage in Cl_2 plasma; (b) SEM photograph of GaAs structure etched in Cl_2 plasma through the As_2S_3:Zn mask.

5. Reactive Ion etching of GaAs in CCL_2F_2 and BCL_3 plasma

CCL_2F_2 is a well known selective etchant of GaAs on AlGaAs. It is used for fabrication of modulation doped FET's and other heterojunction devises. During the etching of GaAs, volatile AsF_xCl_y and $GaCl_x$ products are formed until a barrier layer is created when the GaAs/AlGaAs interface is reached. The barrier layer consists of AlF_3 and $GaCl_xF_y$. This surface layer prevents the etching of aluminum alloys. Fig. 11(a). shows a comparison of the etching rate of GaAs with that of $Al_{0.5}Ga_{0.5}As$ as a function of SB voltage. In this case LPE grows layers were used. Selective etching of $Al_xGa_{1-x}As$ layer occured for $x \geq 3\%$. The RIE was done at low pressure of 6 mTorr. The selectivity of the AlGaAs etching decreased at high SB voltage. At intermediate pressure (30 mTorr) CCl_2F_2 selectively etches GaAs up to 1000 times faster than AlGaAs.

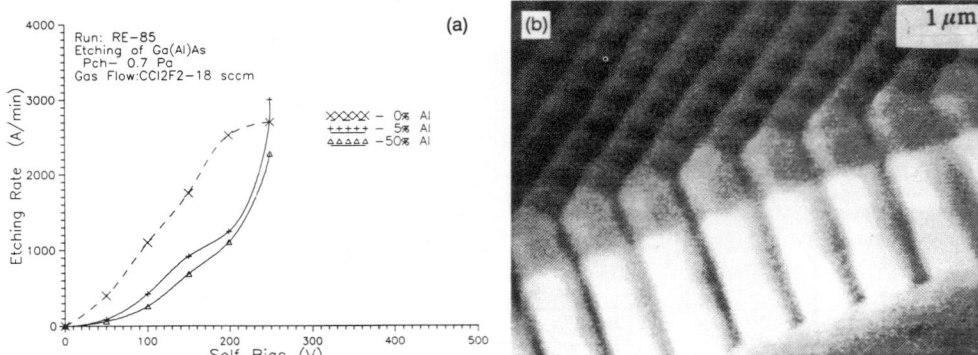

Fig. 11. (a) comparison of the etching rate data on GaAs and $Al_{0.5}Ga_{0.5}As$ film as functions of SB voltage in CCl_2F_2 plasma; (b) SEM photograph of GaAs structure anisotropically etched in CCl_2F_2 plasma. Process parameters: gas flow = 18 sccm, $U_{SB} = -200$ V, pressure in chamber = 0.7 Pa.

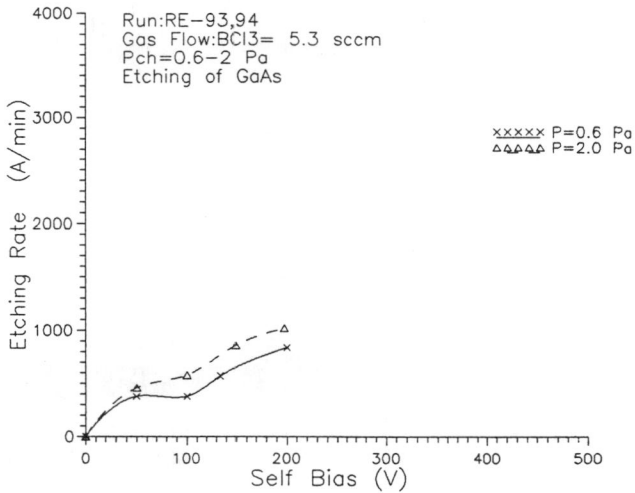

Fig. 12. SB voltage dependence of the etching rate for GaAs in BCl_3 plasma.

Unsaturated CCl_xF_y species can form thin side wall films, which simulate anisotropic etching by ion bombardment. Fig. 11(b) shows a SEM picture of etched GaAs. The etching was done at low pressure (6 mTorr), with gas flow of 18 sccm and 200 V SB voltage. At low pressure the RIE of GaAs in CCl_2F_2 shows acceptable anisotropy and etching rate. At high pressure the unsaturated species can be adsorbed at the surface and polymerize (form thick polymer film). The growth of the polymer film increases with partial pressure of the unsaturated CCl_xF_y species. The polymer film completely terminates the GaAs etching.

BCl_3 can be used for non selective etching of GaAs/AlGaAs structures. Fig. 12 shows the dependence of the etching rate on SB voltage for RIE of GaAs in BCl_3 for two pressures. At low pressure (5 mTorr) the etching rate is intermediate in value and quite anisotropic etching can be achieved at high SB voltage conditions. At intermediate pressures and low SB voltage the etching is isotropic. A very important application of BCL_3 is RIE of thin native oxide films, that are formed on the GaAs, InP and Al surfaces. Furthermore, the plasma decomposition products of BCl_3 scavenge water and oxygen helping to prevent oxide regrowth.[2]

6. RIE of GaAs/Ga(Al)As and InP in Cl_2 plasma

Group III and V elements have usable vapor pressures, especially at elevated temperature.[2] So III–V semiconductors were usually patterned in chloride-containing plasma. The RIE process in pure Cl_2 shows very high etching rate, even at very low pressure. Fig. 13 shows a comparison of etching rates of GaAs/GaAlAs and InP films as functions of SB voltage. The etching rate of InP is lower than that of AlGaAs. Some difference in the etching rates for GaAs and GaAlAs are connected with the presence of water vapor in the chamber. Under intermediate pressures (10 ÷ 50 mTorr) and low SB voltage the etching is crystallographic (Fig. 14(a,b)). V-type groves and mesa-type lasers can be formed. At high SB voltage (> 200 V) conditions the etching rate is very high (> 20000 Å/min) and via-hole connection can be realized. Under very low pressure and high SB voltage (> 150 V) the etching is anisotropic (Fig. 14(c)), especially at high density pattern wafer. Inert gases (Ar or He) are often added in order to stabilize plasma, to enhance the anisotropy or to reduce the etching rate by dilution. High etching rates at low SB voltage can be used for the defect-free RIE of GaAs, because the etching rate of a damaged material can be higher than the rate of the defect diffusion into the wafer from the etched surface.

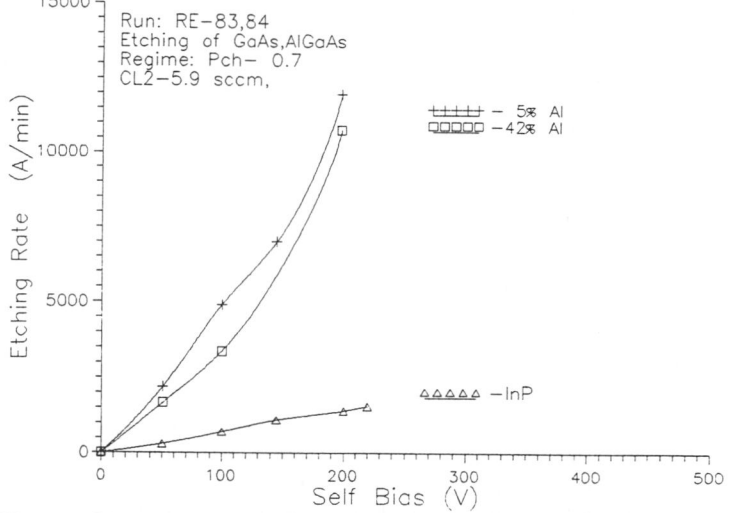

Fig. 13. Comparison of etching rates of GaAs/AlGaAs and InP films as functions of SB voltage in Cl_2 plasma.

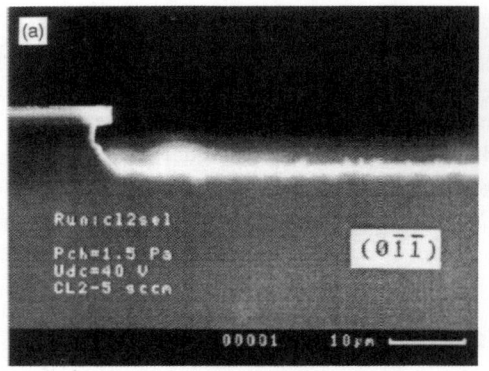
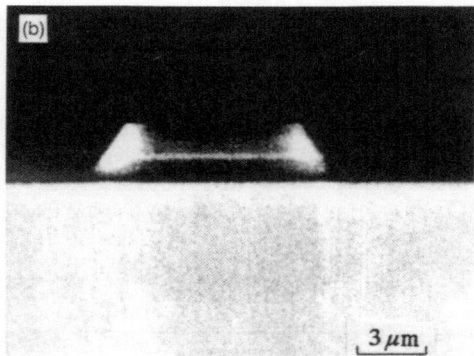
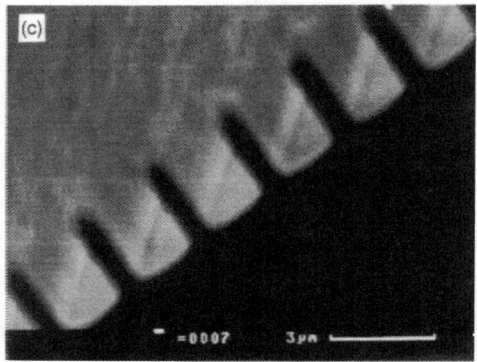

Fig. 14. SEM photograph of GaAs structures etched in Cl_2 plasma. Process parameters: gas flow = 5 sccm, (a) pressure = 1.5 Pa, U_{SB} = −40 V; (b) pressure = 2.0 Pa, U_{SB} = −30 V; (c) pressure = 0.5 Pa, U_{SB} = −200 V.

7. RIE of GaAs/GaAlAs in Cl_2:CH_4+Ar plasma

Very interesting results were obtained when the RIE of GaAs was performed in Cl_2:CH_4 + Ar gas mixture. In this process we used an industrial CH_4 + Ar (1:5) gas mixture. The addition of Ar to Cl_2 helps to stabilize the plasma and to reduce the etching rate. In this case stable SB voltage was less than 5 V. CH_4:H_2 plasma RIE has been successfully employed for metal-organic reactive etching of III–V semiconductors.[7] However, the etching rate for this gas mixture is low. The addition of Cl_2 to the methane containing gas mixture drastically increases the etching rate and can also change the selectivity. Fig. 15(a) shows the dependences of the GaAs etching rates on SB voltage for different CH_4 + Ar flow rates. Fig. 15(b,c) shows a SEM pictures of an 0.25 μm GaAs stripe and a FET structure obtained using a two-layer As_2S_3/AZ-1375 mask, which was etched at 0.7 Pa, CH_4 + Ar flow rate of 15 sccm, Cl_2 flow rate of 5 sccm and SB voltage of 150 V (Fig. 15(b)) and 100 V (Fig. 15(c)). Note the smooth morphology and vertical sidewall without an undercut at the GaAs-mask interface.

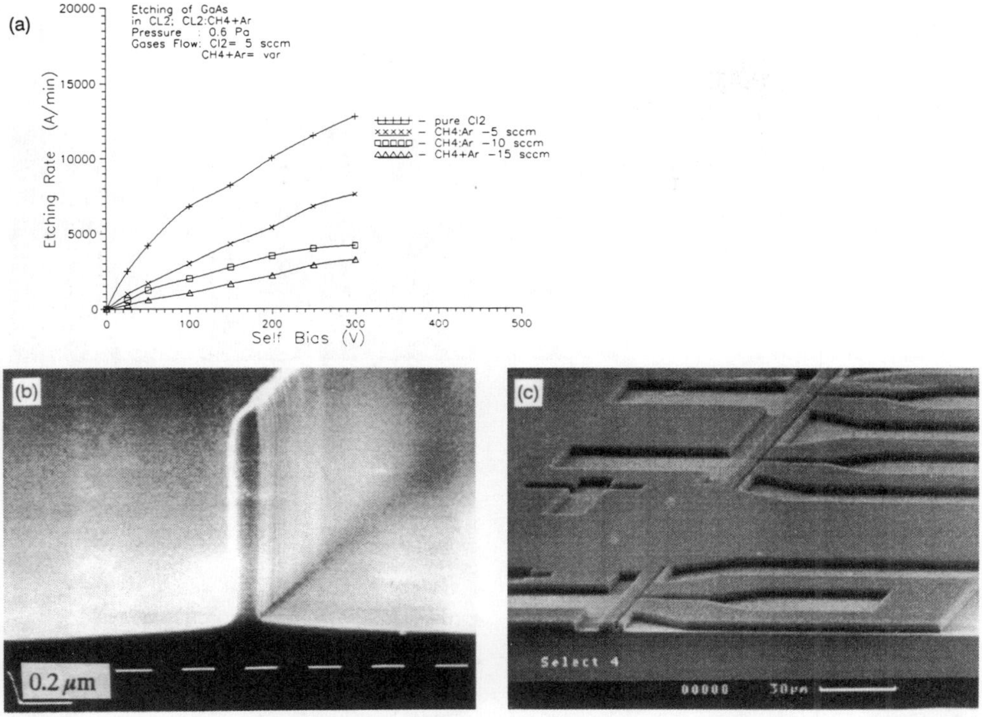

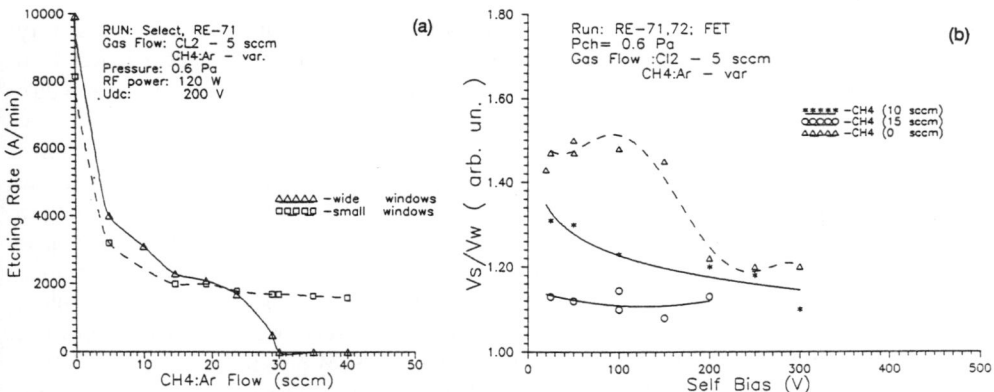

Fig. 15. (a) etching rates vs SB voltage for GaAs etched in $Cl_2:CH_4 + Ar$ plasma for different $CH_4 + Ar$ flow rates. The etching rate data for InP are also shown. SEM photographs of GaAs structures etched at 0.7 Pa in $Cl_2:CH_4+Ar$ (5 sccm : 15 sccm) plasma and SB voltage of: 150 V (b); 100 V (c).

Fig. 16. (a) GaAs etching rates for small and wide pattern windows vs $CH_4:Ar$ flow rate for constant Cl_2 flow rate and $U_{SB} = -200$ V; (b) comparison of GaAs etching rates normalized to wide pattern window as functions of U_{SB} in $Cl_2:CH_4 + Ar$ plasma for different $CH_4 + Ar$ gas flow.

Thin polymer films are formed on the sidewall of a mesa to inhibit the etching of the sidewalls where ions do not impact, or in other words, in this case we have quasi-anisotropic etching

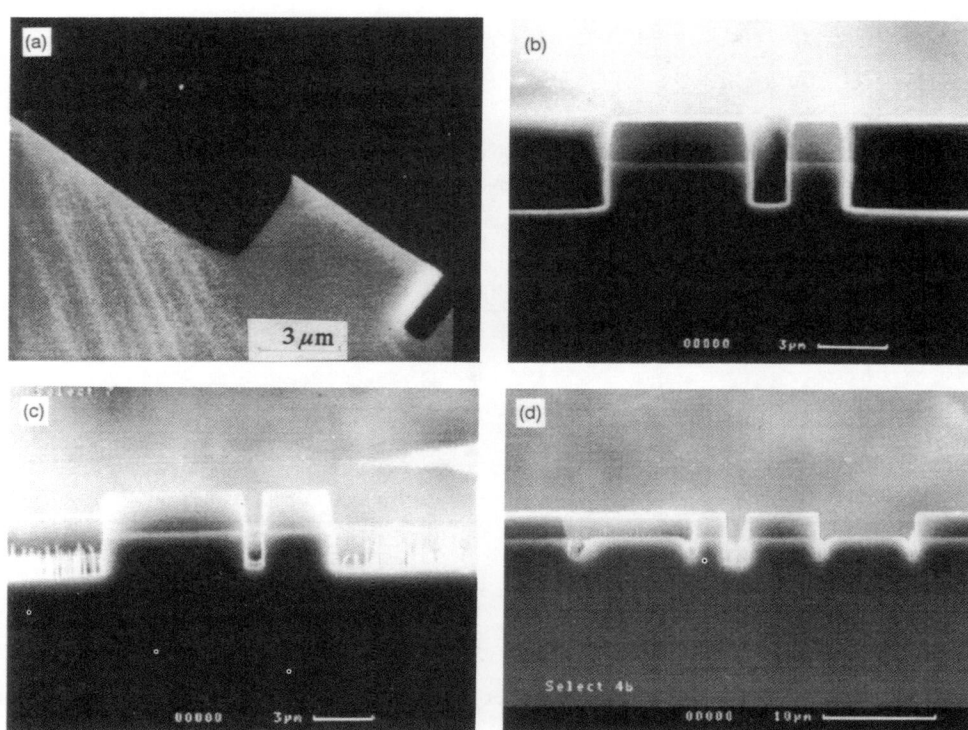

Fig. 17. SEM photographs of GaAs structures etched in $Cl_2:CH_4 + Ar$ ($U_{SB} = -200$ V, pressure = 0.6 Pa, Cl_2 flow rate = 5 sccm) at $CH_4 + Ar$ flow rate of: 0 sccm (a), 15 sccm (b), 25 sccm (c), 32 sccm (d).

condition. The polymer formation process allows to improve microscopic uniformity of RIE, i.e. to solve the problem discussed earlier.[1] It is well known, that etching rates in a RIE process very strongly depends on pattern density and geometry, if the typical size of the pattern is less than 1 μm. For example, smaller windows are etched more slowly than wide windows. Microscopic uniformity can be improved by reducing the pressure. Fig. 16(a) shows GaAs etching rates for small and wide windows versus $CH_4 + Ar$ flow rate for constant Cl_2 flow (5 sccm) at 0.6 Pa and SB voltage of -200 V. As seen from Fig. 16(a,b), the etching rates for small windows (of the size less than 1 μm) at low $CH_4 + Ar$ flow rate (< 10 sccm) are lower than those for wide windows (of the size more than 5 μm). See the SEM picture in Fig. 17(a). The character of etching is crystallographic. When $CH_4 + Ar$ flow rate is increased (> 10 sccm), the etching rate in wide windows decreases, reaching a level which is equivalent to the etching rate in small windows at gas flow rate around 20 sccm. In this condition unsaturated radicals form thin sidewall films and the etching profile of the mesa becomes vertical (see the SEM picture in Fig. 17(b)). When $CH_4 + Ar$ mixture flow rate reaches 25 sccm level the polymer deposition process on the free of mask surface of GaAs becomes stronger, and island-type masking of the surface occurs (see the SEM picture in Fig. 17(c)). If the methane-argon flow rates are equal approximately to the 30 sccm, polymer deposition process "takes over" and the etching process is completely stopped in wide windows (see the SEM picture in Fig. 17(d)). At the same conditions the etching rate in narrow windows and near the edge of the mask is practically the same as at 15 sccm $CH_4 +$

Ar flow rate. This fact can be explained by the following way. Ion reflection from the vertical mask edge results in increased ion density and increased etching rate of the polymer films on GaAs near the edge of the mask. In Fig. 18(a) a comparison of the etching rate of GaAs with that of AlGaAs as functions of SB voltage is shown. The etching was made under optimal Cl_2 (5 sccm) and $CH_4 + Ar$ (14.6 sccm) flow conditions. The etched samples were $\cong 0.3 \, cm^2$ LPE grown (2 μm in thick) n-type GaAs or $Al_{0.42}Ga_{0.58}As$ films on GaAs substrate. The results shown in Fig. 18(a) indicate that as SB voltage is less than 100 V, the GaAs etching rate is higher as compared with AlGaAs. This is typical for $CH_4:H_2$ gas MORIE.[7] At higher SB voltages (> 120 V) the AlGaAs etching rate is higher than that for GaAs (Fig. 18(a)). Fig. 18(b) shows a SEM picture of etched GaAs/AlGaAs DHS laser mesa structure. The RIE was performed at 150 V. Note the same undercut of AlGaAs layers at the sidewalls of the mesa. When the RIE was performed at SB voltage $120 > U_{SB} > 80$ volts nonselective etching of AlGaAs/GaAs occured. Fig. 18(c) shows a SEM picture of a RIE etched multi-stripe single-quantum-well laser structure. Plasma conditions were the following: 0.7 Pa; $U_{SB} = -100$ V; $Cl_2 = 5$ sccm; $CH_4 + Ar = 15$ sccm.

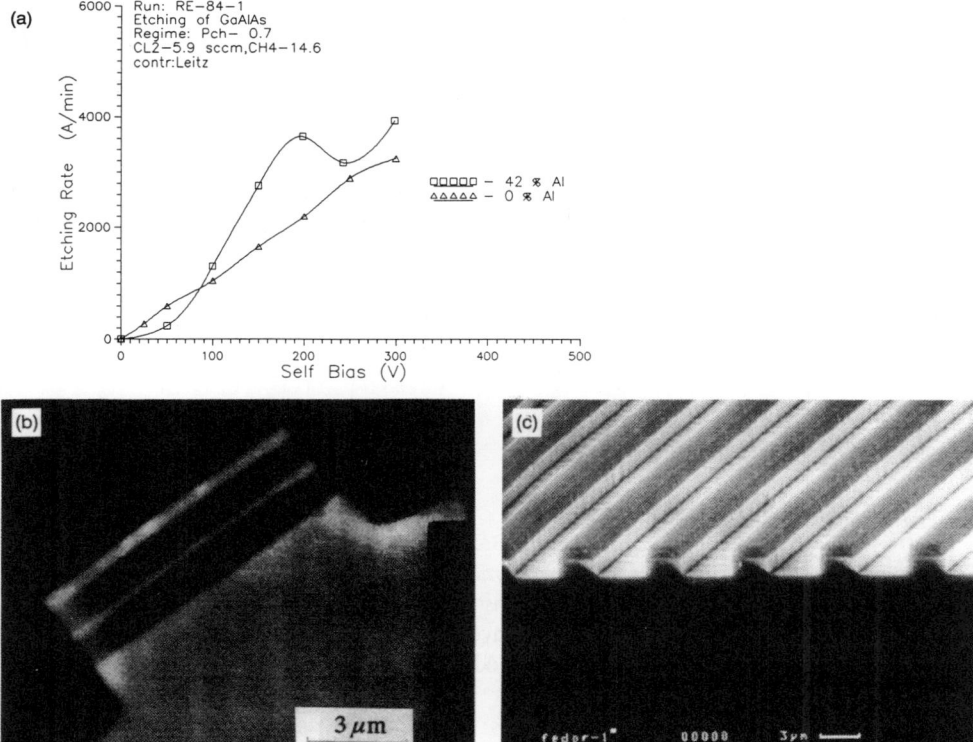

Fig. 18. (a) comparison of etching rate of GaAs with $Al_{0.42}Ga_{0.58}As$ as a function of SB voltage; (b) SEM photograph of GaAs/AlGaAs DH laser mesa structure etched in $Cl_2:CH_4 + Ar$ (6:15 sccm), pressure = 0.7 Pa, $U_{SB} = -150$ V; (c) SEM photograph of GaAs/AlGaAs SQW LOC multi-stripe laser structure etched at the same conditions, but $U_{SB} = -100$ V.

8. Study of plasma etching damage in GaAs

The quality of the surface obtained after RIE of GaAs is of primary importance in electronic and optoelectronic applications, such as recessed channel etched FET, etched stripes and mirrors for diode lasers, quantum wire and quantum box device formation. The RIE damage of material is associated with ion bombardment effects. Different methods can be applied for investigation of intrinsic properties of the GaAs surface: Raman scattering,[8,9] photoluminescense,[13] x ray photoelectron spectroscopy,[10] SIMS and Auger microanalysis[11], and others.

8.1. Raman scattering study of damage in GaAs

Raman spectra were exited by the Ar-ion laser (5145 Å line, 15 mW optical power) in the backscattering geometry, where incident and scattered light propagate along the [100] axis of a GaAs sample. Using of the 5145 Å laser line allowed us to probe the substrate within the penetration depth of approximately 1000 Å, where most of the RIE associated damage is expected to take place. n^+-type unpatterned GaAs samples (Sn-doped, $n = 1 \div 2 \times 10^{18} cm^{-3}$) were used to study structural damage and free electron gas condition after the reactive ion etching in CF_2Cl_2, Cl_2, Cl_2:CH_4 + Ar plasmas. Initial GaAs samples were chemically etched to remove about 30 μm of GaAs. In order to remove the same quantity of the material (100 Å $\div$ 3 μm), the samples were etched in different plasma for different time and then removed for characterization. In backscattering geometry from a (100) surface of a zinc-blende structure crystal, such as GaAs, only the LO phonon mode is allowed.[12] When the heavily doped GaAs crystals are sampled by laser beam in backscattering geometry, the Raman spectrum is a superimposition of the unscreened LO phonon mode originating from the surface depletion region and the coupled plasmon-LO phonon modes from the underlying region with free carriers[9] (see Fig. 19). The ratio of intensities of the LO phonon and the low-frequency coupled mode L^- gives a measure of the width of the depletion layer. For unprocessed wafer this intensity ratio, expressed in terms of the absorption coefficient of the laser beam in the sample α and the depletion layer width d_0, is given by

$$I_0 = \frac{I(LO)}{I(L^-)} = A[\exp(2\alpha d_0) - 1] \tag{1}$$

$\alpha = 10^5 cm^{-1}$—the absorption coefficient for GaAs; $d_0 \cong 240$ Å—depletion layer width of the unprocessed sample, evaluated from the free carrier concentration given by the L^+ peak position and the GaAs barrier height (0.8 eV); A—a constant parameter for given sample and its value is calculated from (1). At the second stage we have to measure ratio of the intensities of the LO phonon mode and the low-frequency coupled mode L^- for RIE processed samples I_1

$$I_1 = \frac{I(LO)}{I(L^-)} \tag{2}$$

Finally, for processed samples the depletion layer width can be calculated as

$$d_1 = \ln(\frac{I_1}{A} + 1) \cdot (2\alpha)^{-1} \tag{3}$$

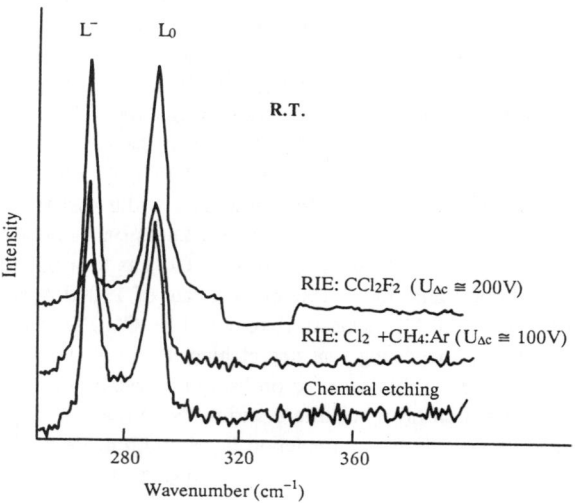

Fig. 19. Typical Raman scattering spectrum for reference material (chemically etched GaAs) and samples subjected to: RIE in $Cl_2:CH_4+$ Ar plasma (U_{SB} = −100 V, gas flow = 5:15 sccm, 10 min, pressure = 0.7 Pa); RIE in CCl_2F_2 plasma (U_{SB} = 200 V, gas flow = 20 sccm, 20 min, pressure = 0.7 Pa).

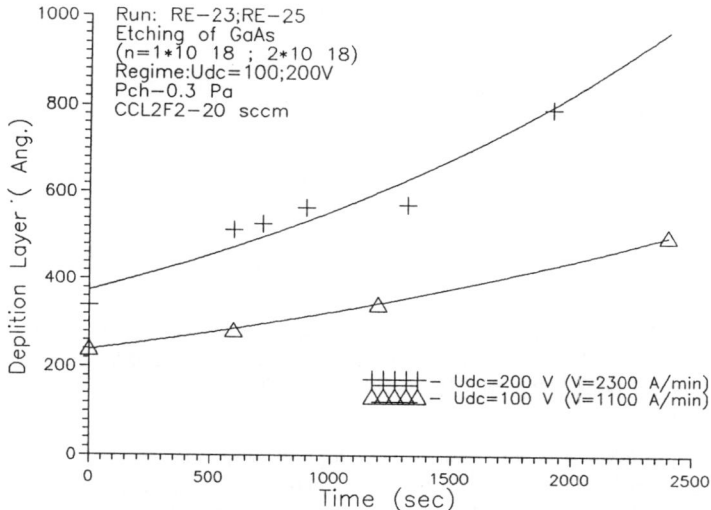

Fig. 20. The dependence of depletion depth on etching time for GaAs ($n = 1 \times 10^{18}$ cm^{-3}; $n = 2 \times 10^{18}$ cm^{-3}) samples, etched in CCl_2F_2 plasma at: U_{SB} = 100; 200 V; the etching rates were 1100 Å/min and 2300 Å/min, respectively

The Raman spectra of the RIE samples are compared to the reference unprocessed sample in Fig. 19. We observe peaks at 269 and 292 cm^{-1}, corresponding to the coupled plasmon-LO phonon (L^-) and the longitudinal optical phonon (LO). Within the experimental resolution, the

LO phonon linewidth is the same for all of the samples, indicating that the surface disorder is relatively small, but nevertheless still sensitively probed by the measured changes in the depletion layer width. Fig. 20 shows the dependence of depletion layer width on etching time for GaAs ($n = 1 \times 10^{18}$ cm^{-3}; $n = 2 \times 10^{18}$ cm^{-3}) samples, etched in CCl$_2$F$_2$ at 100 and 200 V of SB voltage. At this condition of RIE, the etching rate for GaAs was 1100 Å/min ($U_{SB} = 100$ V) and 2300 Å/min ($U_{SB} = 200$ V). The plot in Fig. 20 shows that the depletion layer width increases with etching time (etching depth) even at low energy RIE (100 eV). When a 4.5 μm layer of GaAs was removed in CCl$_2$F$_2$ plasma, the depletion layer width was approximately doubled. Fig. 21 shows the dependence of the depletion layer thickness on etching time for GaAs ($n \cong 1.8 \times 10^{18}$ cm^{-1}), etched in Cl$_2$ plasma at 0.6 Pa, 5.6 sccm Cl$_2$ gas flow and 200 V SB voltage (the etching rate of 10000 Å/min) and 300 V (the etching rate of 13000 Å/min). These data indicate, that when 7.5 μm ($U_{SB} = 200$ V) of GaAs is removed by Cl$_2$ RIE the depletion layer width is reduced by a factor of 1.5, suggesting that dry etching does not introduce any damage and/or impurities at the GaAs surface. In this case we probably observe an effect of surface passivation after RIE, because the depletion layer width are shrink. Note that for $U_{SB} = 300$ V RIE when the etching rate of GaAs is very high, the depletion layer width drops to 130 Å (Fig. 21) during the first few minutes of the RIE and after that monotonically recovers to its previous level.

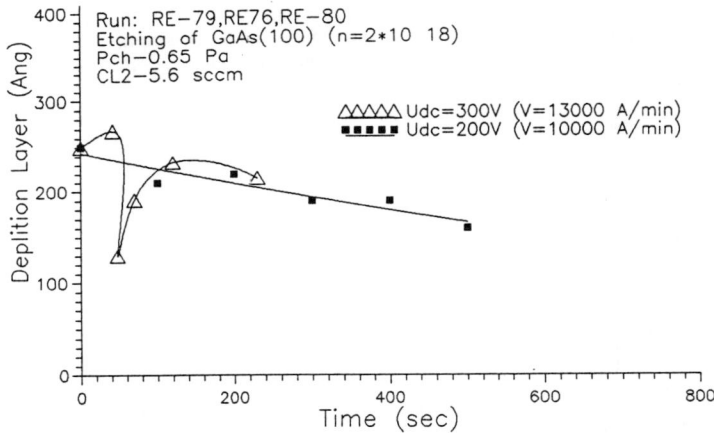

Fig. 21. The dependence of depletion depth on etching time for GaAs ($n = 1.8 \times 10^{18}$ cm^{-3}), etched in Cl$_2$ plasma at: $U_{SB} = 200$ V (etching rate = 10000 Å/min) and $U_{SB} = 300$ V (etching rate = 13000 Å/min).

RIE by Cl$_2$ plasma is characterized by very high etching rate, so the rate of defect generation in GaAs by ion bombardment can be lower than the etching rate. This effect combained with the surface passivation can result in depletion layer width reduction. In general, this problem requires additional investigations. In Fig. 22 the depletion layer width versus RIE time is plotted when GaAs ($n = 1.8 \times 10^{18}$ cm^{-3}) was etched in Cl$_2$:CH$_4$ + Ar plasma at 0.6 Pa, 5.6 sccm Cl$_2$ and 14.6 sccm CH$_4$ + Ar gas flow rates and 100 and 200 V of SB voltage. In this case the etching rate was 1000 Å/min ($U_{SB} = 100$ V) and 2200 Å/min ($U_{SB} = 200$ V), or practically the same as for the CCl$_2$F$_2$ RIE of GaAs. The depletion layer for the methane-argon-chlorine RIE at $U_{SB} = 200$ V was slightly wider than that of $U_{SB} = 100$ V and the same as for unprocessed GaAs wafer, in contrast to the CCl$_2$F$_2$ RIE of GaAs (see Fig. 20).

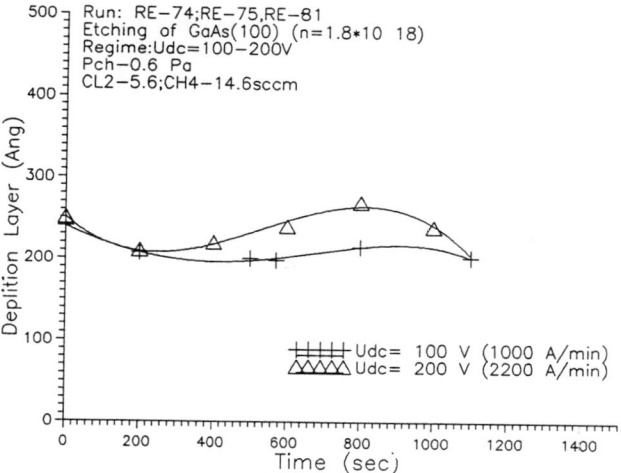

Fig. 22. The dependence of depletion depth on etching time for GaAs ($n = 1.8 \times 10^{18}$ cm^{-3}) samples, etched in Cl$_2$:CH$_4$ + Ar plasma at: $U_{SB} = 100$ V (etching rate = 1000 Å/min); $U_{SB} = 200$ V (etching rate = 2200 Å/min).

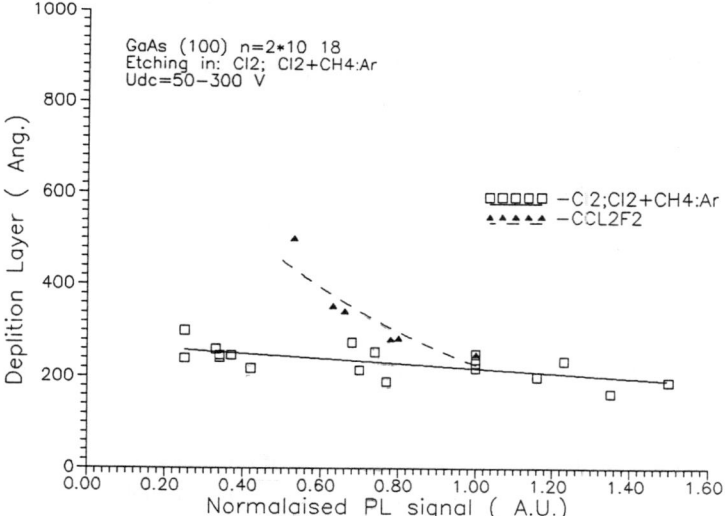

Fig. 23. Correlation between PL signal normalized to an unprocessed sample and depletion layer width for GaAs (2×10^{18} cm^{-3}) etched in Cl$_2$:Cl$_2$:CH$_4$ + Ar and CCl$_2$F$_2$ plasmas.

We also collect PL data for all the samples studied by Raman probe technique. PL was excited by the same green line of Ar$^+$-ion laser (15 mW) at room temperature. Fig. 23 shows the correlation between the PL signal normalized to that for the unprocessed sample and the depletion layer width (derived from our Raman scattering studies) for GaAs ($n = 2 \times 10^{18}$ cm^{-3}) etched in Cl$_2$ or Cl$_2$:CH$_4$ + Ar plasma at $U_{SB} = 50 \div 300$ V. The plot shows that high luminescence efficiency corresponds to a narrow depletion layer. Note that for GaAs samples etched in CCl$_2$F$_2$ plasma the PL—depletion layer width correlation plot has a different slope (triangles in Fig. 23) and show a stronger dependence of the PL signal on depletion layer width. In order to check the effect of sample processing on the surface quality of a real semiconductor structure we investigated the PL spectrum of MQW, etched step by step in different plasma under low SB condition.

8.2. The effect of low energy RIE on photoluminescence of MQW structure

The GaAs/AlGaAs MQW structures were made from a material grown by MOCVD on an i-GaAs substrate, and consisted of the following undoped GaAs and $Al_{0.28}Ga_{0.72}As$ layers in order of growth: 0.5 μm GaAs, 0.3 μm AlGaAs, 17 nm GaAs, 30 nm AlGaAs, 9.6 nm GaAs, 30 nm AlGaAs, 5 nm GaAs, 30 nm AlGaAs, 2.5 nm GaAs and 30 nm AlGaAs. The 0.3 μm AlGaAs layer was included to inhibit diffusion of photoexited carriers from the 0.5 μm GaAs buffer layer and the substrate to the quantum wells (see Fig. 24(a)). The wells and the barriers were verified by transmission electron microscopy and by analysing of spectral positions of peaks in the PL spectrum. The structure was patterned by usual photolithography technique in order to perform 1 × 1 mm opened windows. After that, RIE was performed through these windows. One part of MQW structure was etched in O_2 plasma, another part was etched in Cl_2:CH_4 + Ar plasma. The O_2 plasma RIE was performed at $U_{SB} = 200$ V with 0.6 Pa pressure and 20 sccm gas flow. The Cl_2:CH_4+Ar plasma RIE was performed at $U_{SB} = 25$ V with 0.6 Pa pressure and gas flow of 5 sccm for Cl_2 and 15 sccm for CH_4+Ar. A part of structures protected by a photoresist (unprocessed in RIE) was used as a reference point in PL study. The PL characterization was performed at 77 K; for MQW structure excitation we used 514.5 nm 50 mW line of Ar-ion laser. Fig. 24(b) shows a comparison of PL spectra of unprocessed and O_2-plasma etched structures. For the unprocessed structure one can see 4 PL peaks: the short wavelength peak corresponds to photon emission from 2.5 nm QW, and long wavelength peak corresponds to emission from the 17 nm QW. In oxygen plasma a 62 nm layer of the structure was removed and the etching was stopped inside the 5 nm QW. As is seen from Fig. 24(b), only one peak exists in PL spectrum and the amplitude of this signal is lower then that for the unprocessed structure. This result indicates that oxygen plasma RIE introduces a damage which reduces the PL efficiency of the QW situated at 30 and 65 nm under the etched surface. GaAs and AlGaAs oxygen-exposed surface is possessed of a large density of extrinsic states near the middle of the band gap which effectively pin the Fermi level. This midgap pinning is associated with trapping centers, whith can give rise to high nonradiative recombination rate reducing PL efficiency. This can limit the performance of minority-carrier devices, such as light-emitting diodes.[13] Fig. 24(c) shows a comparison of PL spectra from unprocessed and Cl_2:CH_4+Ar low-energy plasma RIE processed parts of the MQW structure. In this case 10 nm and 45 nm layers of the structure were etched. When a 10 nm layer of AlGaAs barrier has been removed the PL intensity from the 2.5 nm QW is reduced because of the surface recombination processes (the QW is situated closer to the surface inside the depletion layer). In contrast, the PL efficiency of the other 5, 9.6 and 17 nm wide QW increases, because the carrier recombination probability increases in these wells. The thickness of AlGaAs layers is reduced and optical losses caused by these layers are reduced, so more carriers can be generated in AlGaAs barrier layers around the 5, 9.6 and 17 nm QW's. When 45 nm layer of the structure has been etched at the same conditions, the PL efficiency of the MQW structure continues to increase (Fig. 24(c)). This result indicates that RIE in Cl_2:CH_4 + Ar plasma can provide high-quality defectless dry etching of real MQW structures and can be applied, for example, for quantum wire and quantum box structure formation.

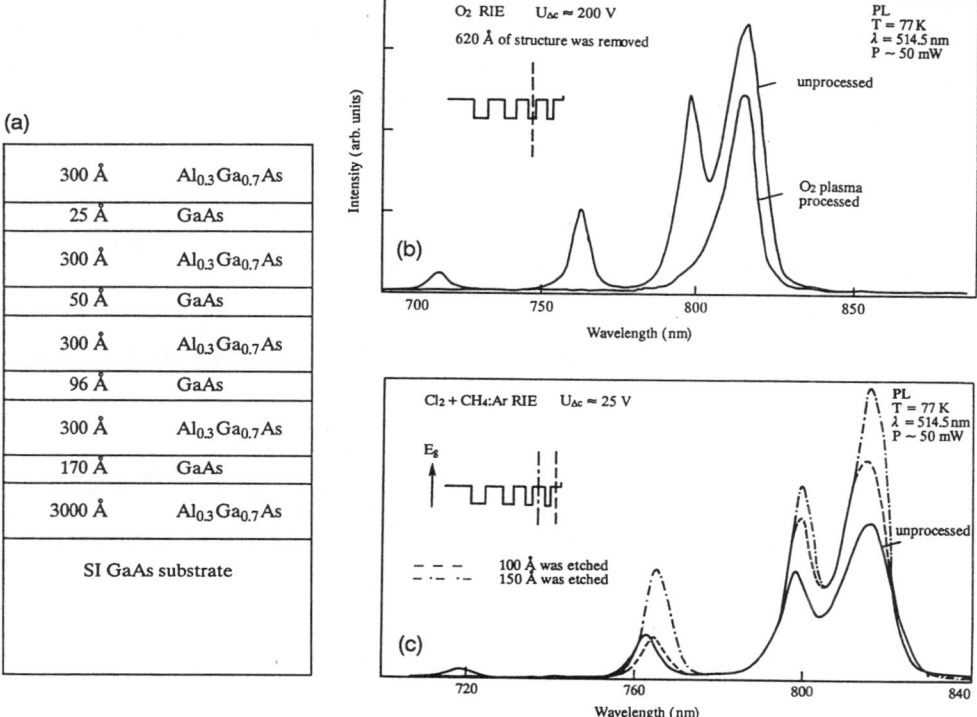

Fig. 24. (a) diagram (not to scale) showing a GaAs/Al$_{0.3}$Ga$_{0.7}$As MQW structure, grown on SI GaAs substrate by MOCVD; (b) representative PL spectra obtained at 77 K with excitation at 514.5 nm and 50 mW for the GaAs/AlGaAs MQW sample: unprocessed and when 62 nm of structure was removed in O$_2$ plasma. RIE process parameters: $U_{SB} = 200$ V, gas flow = 20 sccm, pressure = 0.7 Pa; (c) PL spectra obtained at the same conditions for the same MQW sample: unprocessed and when 10 nm and 42 nm of structure were removed in Cl$_2$:CH$_4$ + Ar plasma. RIE process parameters: $U_{SB} = 25$ V; Cl$_2$ flow = 5 sccm; CH$_4$ + Ar flow = 14 sccm; pressure = 0.7 Pa.

9. Conclusion

The work presented above provides a review consideration of the problems related to choosing of a suitable mask, etch chemistries and etching process parameters for submicrone elements' formation and defect-free RIE of GaAs/AlGaAs and InP structures. It is shown that the multilayer systems containing As$_2$S$_3$ can be used as a high resolution self-removable mask for RIE. The data of Raman scattering and PL studies of RIE of GaAs are presented. Low energy defectless RIE of the MQW AlGaAs/GaAs structure has been realized.

Acknowledgments

The author thank Drs. V. Smirnitsky, A. Il'menkov S. Nesterov, M. Kulagina, A. Kolobov and P. Kopiev for process support, Drs. I. Kyzmin and B. Yavich who provided the growth of

MOCVD structures. Process control software creation was provided by N. Belenky, A. Titov and P. Il'menkov. Drs. V. Melekhov and Yu. Akulova are acnowledged for PL and Raman microprobe data collection. Special acknoledgements are adressed to Drs. S. Gurevich, A. Dul'kin and prof. J. Canteloup (from "Sofie Instrument") for helpful discussions and to E. Solov'eva and N. Vsesvetsky for help in paper preparation.

References

1. K.P. Giapis, G.R. Scheller, R.A. Gottscho, W.S. Hobson, and Y.H. Lee 1990 *Appl. Phys. Lett.* **57** 983
2. D.E. Ibbotson and D.L. Flamm 1988 *Sol. State Tech.* **October** 77
3. A. Mitchel and R.A. Gottscho 1990 *J. Vac. Sci. Technol.* **A8** (3) 1712
4. E. Reichmanis, G. Smolinsky, and C.W. Wilkins, Jr. 1985 *Solid State Technol.* **August** 130
5. B.T. Kolomiets, V.M. Lubin, and V.P. Shilo 1978 *Phyzika i Khimiya Stekla* **4** (3) 351
6. Zh.I. Alferov, S.A. Gurevich, S.Yu. Karpov, E.L. Portnoi, and F.N. Timofeev 1987 *IEEE J. Quantum Electron.* **23** (6) 869
7. V.J. Law, G.A.C. Jones, and M. Tewordt 1990 *Semicond. Sci. Technol.* **5** 1001
8. D. Kirilov, C.B. Cooper,III, and R.A. Powell 1986 *J. Vac. Sci. Technol.* **B4** (6) 1316
9. B. Roughani, H.E. Jackson, J.J. Jbara *et al.* 1989 *IEEE J. Quantum Electron.* **25** (5) 1003
10. G.S. Oehrlein, K.K. Chan, M.A. Jaso, and G.W. Rubloff 1989 *J. Vac. Sci. Technol.* 1030
11. M. Uchida, S. Ishikawa, N. Takado, and K. Asakawa 1988 *IEEE J. Quantum Electron.* **24** (11) 2170
12. W. Hayes and R. Loudon *Scattering of Light by Crystals, New York: Wiley 1978* p.147
13. S.R. Andrews, H. Arnot, P.K. Rees, T.M. Kerr, and S.P. Beaumont 1990 *J. Appl. Phys.* **67** (7) 3472

Monolithic multiple wavelength tunable vertical cavity surface emitting laser array

C.J. Chang-Hasnain, M.W. Maeda, J.P. Harbison, and L.T. Florez
Bellcore, 331 Newman Springs Road, Red Bank, NJ 07701 USA

Abstract. The wavelength multiplexing and tuning capabilities are highly important for applications in optical interconnects and optical communications using wavelength division multiplexing (WDM). In this paper, we describe novel methods and experimental results to achieve these functions in vertical cavity surface emitting lasers for the first time. We achieved a 2-dimensional surface emitting laser array emitting 140 unique, nonredundant, uniformly separated, single-mode wavelengths. The wavelength separation between neighboring lasers is as small as 0.3 nm. A large total wavelength span of 43 nm is obtained without compromising the performance of the lasers. All 140 lasers have nearly the same threshold currents, voltages, and resistances. The techniques used here are generic and can be readily extended to both longer and shorter wavelength lasers. We also report the first WDM system experiment using part of this laser array. A bit-error-ratio of 10^{-9} at 155 Mb/s was obtained with simultaneous operation of four lasers at a wavelength separation of 1.5 nm. Negligible optical and electrical crosstalk was observed between the lasers. In addition, we report the first continuously wavelength-tunable vertical cavity surface emitting laser. The wavelength tuning is achieved by pumping the tuning current through the n-doped Bragg reflectors and the n^+-GaAs substrate with an additional electrode placed near the laser junction. A tuning range of 18 Å (equivalent to 540 GHz) is obtained with no mode hopping. A $4 \sim 6$ Å blue-shift is achieved with approximately 14 mA tuning current. The laser output power remains nearly constant with the tuning current throughout the tuning range. The tuning speed is determined to be approximately $300 \sim 400$ ns using time-resolved spectrum measurements. These results and the techniques used have opened up a new avenue for applications and system architectures using laser wavelength as an additional degree of freedom.

1. Introduction

Semiconductor diode lasers emitting normal to the substrate plane, known as surface emitting lasers, are extremely attractive for addressing a range of applications. This is because they facilitate wafer-scale fabrication and testing of potentially cost-effective 2-dimensional (2D) laser arrays. At present there are two orthogonal approaches aimed at realizing such lasers. The first represents an extension of the existing technology for semiconductor edge-emitting lasers that uses a grating[1] or tilted mirror[2] to vertically couple the light out. The second is a radical departure from the traditional semiconductor laser design. In this case, highly reflective mirrors[3] are used to clad the active region resulting in a vertical cavity that produces an output beam propagating normal to the substrate surface.

Recently, there have been many advances[4-11] on vertical cavity surface emitting lasers (VCSEL). The advantages of using a vertical cavity are numerous. The active volume of VCSEL can be made so small that high packing density, low threshold current lasers can be obtained.[5] The integration of VCSEL with electronic or other optical devices is also expected to be easier due to the small laser sizes. In addition, VCSEL's emit circular low divergence output beams,[6] which make coupling with optical fibers[7] and even bulk optics very easy and efficient.

For local area communications, long distance optical fiber communications, and optical interconnects from board-to-board level to computer-to-computer level, architectures using wavelength division multiplexing (WDM) provides an extra dimension of freedom—wavelength and, therefore, more flexibilities and higher capacities. An important optical source for WDM systems is an ensemble of single-wavelength lasers emitting unique wavelengths with uniform wavelength spacings. Such an ensemble can be made of hybrid lasers which must be individually packaged and carefully wavelength-selected. Hence, this approach tends to be very costly and the wavelengths not easily reproducible. A monolithic multiple wavelength laser array is thus highly desirable. Another key functionality for systems using WDM is a wavelength tunable laser or laser array, particularly if the tuning mechanism can be implemented on a multiple wavelength laser array. The wavelength tuning can provide the locking and stabilization of each transmission wavelength. Moreover, it introduces a highly desirable capability to a network—reconfigurable interconnections.

Both multiple wavelength laser array and wavelength tunable lasers have been previously reported for edge emitting laser structures. In this paper, we present novel techniques and experimental results to achieve these operations in surface emitting laser structures. In Sec. 2, the realization of a multiple wavelength laser array is described, which is followed by Sec. 3 with the description of a novel wavelength tunable laser. Some concluding remarks are discussed in Sec. 4.

2. Realization of a multiple wavelength surface emitting laser array

The techniques used to obtain edge-emitting, multiple wavelength laser arrays can be put into 2 categories based on the laser structures. For quantum-well Fabry–Perot (FP) lasers, one technique is to vary laser geometry and cavity loss so that lasing transitions occur from different quantum-well subbands, and thus different wavelengths.[12] Using this technique, the number of different wavelengths reported is only 2. Although the number can probably improved to 3 or 4, the wavelength separation $\Delta\lambda$ will remain large (20 ~ 30 nm) and hard to control, and the laser uniformity remain poor. Another technique is to use laser-induced desorption to taper the quantum well layer thickness.[13] This technique also suffers from $\Delta\lambda$ being hard to control and poor laser uniformity. For DFB (distributed feedback) and DBR (distributed Bragg reflector) lasers, the optical grating period can be varied to achieve multiple wavelength emission. However, high precision control of the grating periods is required. Previously, x ray[14] and e-beam lithography[15] were used. The smallest $\Delta\lambda$ reported was ~ 0.7 nm, which is very close to the theoretical limit. The processing for these laser arrays is complicated and the yield tends to decline drastically with the number of lasers. The maximum number of wavelengths that can be integrated in a one-dimensional array is, thus, limited.

In this section, we describe a VCSEL array which, in contrast to the above mentioned schemes, can provide hundreds of independent wavelengths through control during the wafer growth step. The wavelength spacing is uniform and is as small as 0.3 nm. It can be, in principle, made smaller than 0.1 nm or as large as several nanometers. Using this novel design, not only a larger number of controllable wavelengths can be obtained, the processing is no more complicated than what is required for a single VCSEL. A WDM-based system experiment using part of the laser array is also described here, which is the first system experiment using a VCSEL array. The results indicate negligible optical or electrical crosstalk between the lasers.

This section is organized as follows. We first describe the idea and the principle of operation. Next, we discuss the implementation of the idea, namely the actual device design and fabrication. We show the experimental results of the laser array and describe a WDM system experiment demonstrated using part of the array. Some design variations of our techniques and potential applications enabled by such a multiple wavelength laser array are then discussed.

2.1. Principle of array design

With some calculations on the reflectivity and transmission spectra of a standard VCSEL design,[6,8] some important and unique properties of VCSEL became clear to us. This understanding then led to the basic idea for the realization of multiple wavelength arrays. The VCSEL used for our calculations include 14 pairs and 20.5 pairs of alternating quarter-wave stacks of GaAs/AlAs distributed Bragg reflectors as the top and bottom mirrors, respectively, and InGaAs strained quantum well active region placed at the center of a one-wave $Al_{0.5}Ga_{0.5}As$ spacer. The entire structure was designed for 980 nm. (The details of the heterostructure used experimentally will be described in Sec. 2.2.) The calculation shows that there exists only one Fabry–Perot mode within the mirror bandwidth due to the ultrashort cavity. The mode, which is the VCSEL lasing wavelength, depends critically on the total cavity length. Fig. 1 shows the calculated Fabry–Perot mode wavelength of the VCSEL as a function of layer thickness variation in 2 pairs of DBR which are closest to the active layer. The wavelength varies approximately linearly and monotonically with the layer thickness. With the 2-pair being 10% thicker, the wavelength red shifts 15 nm. As large as 50 nm in wavelength difference can be obtained with only ±15% thickness variation. It is worth to mention that the effect of the wavelength shift is more pronounced with the pairs having thickness variation closer to the active region. Secondly, the DBRs have wide reflectivity bandwidths, ~ 100 nm if taken at 95% of the peak value. Within the bandwidth, the reflectivity spectrum is quite flat. The reflectivity spectrum and its peak value are very insensitive to small thickness variation in the Bragg reflectors. Hence, the required high reflectivity in the mirrors for a VCSEL is not compromised with small variations in Bragg reflector thickness. Finally, the optical field peaks at the quantum wells resulting in a desirable large overlap of gain and optical field. This overlap remains large for the case when the VCSEL cavity length is varied slightly. Thus, it is clear that the VCSEL lasing wavelength can be tailored to either longer or shorter wavelength with a small variation in cavity thickness without compromising the lasing characteristics.

Since the VCSEL wavelength depends sensitively on cavity length, a thickness gradient across the wafer can translate into a wavelength gradient of the lasers fabricated across the wafer. Fig. 2 shows the schematic of a 3-element laser array based on this idea. The details on the laser structure and fabrication will be described in the next section, this figure is shown here to illustrates the idea clearly. The laser structure consists of a top and bottom DBR and a center one-wave spacer, in the middle of which are the active layers. As shown, the lasers are defined by proton implant. The spatial thickness gradient is implemented in two layers closest to the central spacer. Note the amount of thickness variation shown here is highly exaggerated. The lasing wavelengths for these three lasers are thus tailored according to the thickness variation.

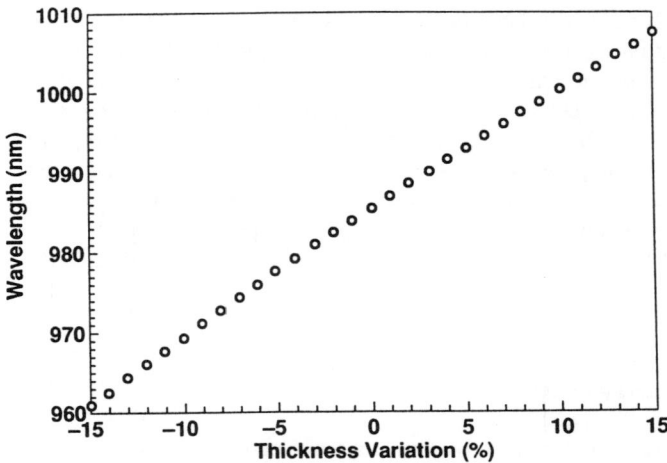

Fig. 1. The calculated Fabry–Perot mode wavelength of the VCSEL as a function of layer thickness variation in 2 pairs of 20 pair bottom DBR.

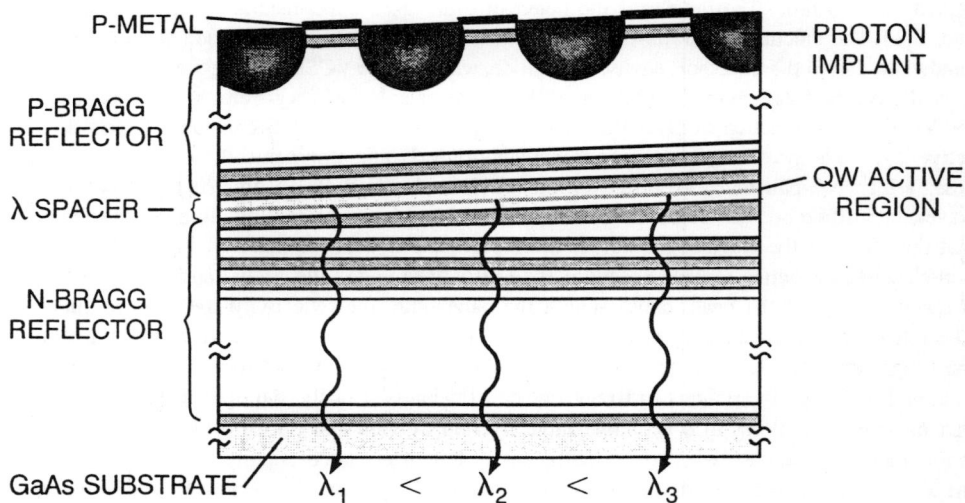

Fig. 2. Schematic of an array consisting of three VCSEL's as defined by proton implant. Note the amount of thickness variation in the chirped layers is highly exaggerated. Also, the distance between the lasers is typically much larger than the laser dimension.

One way to create such thickness gradient is to simply keep the wafer stationary during part of the molecular beam epitaxy (MBE) growth. The thickness variation originates from the fact that the atomic sources in an MBE system are incident to the wafer at an angle off normal ($\sim 33°$) for the Varian Gen II system we used) and hence the number of atoms arriving at the wafer varies monotonically in the direction parallel to the plane of incidence of the sources. Fig. 3 shows schematically the arrangement of the atomic sources in an MBE system. Since the MBE material is grown in an As rich environment, the thickness of the MBE growth is determined by the number of group III sources that arrive at the wafer. Thus, the directions of thickness variation for GaAs and AlAs layers are parallel to the directions of the Ga and Al sources, respectively. Therefore, we can obtain the desired small but definite thickness variation across the wafer by rotating the wafer for uniformity during the MBE growth of the VCSEL structure **except** for two pairs of AlAs and GaAs DBR layers during whose growth the wafer is kept stationary. Since the Ga and Al sources are placed next to each other in our MBE system, the direction of the resulting cavity thickness variation runs parallel to the line intersecting the directions of the two sources.

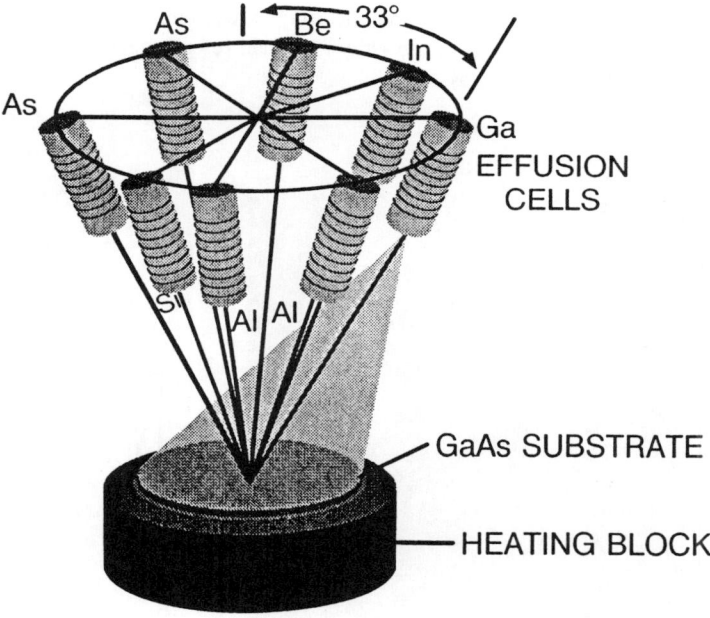

Fig. 3. The arrangement of the atomic sources in an MBE system. For ease of illustration, the schematic is rotated by 90° i.e. the substrate is actually held vertically rather than lying horizontally.

Although the wafer used here is grown with thickness variation only in one direction, $\vec{t}$, nevertheless, a 2D laser array having no redundant wavelengths and a well-defined wavelength variation among the lasers can be obtained if the direction $\vec{t}$ does not coincide with the array axes. We have shown that the laser wavelength depends linearly on the cavity thickness in Fig. 1. In addition, over a small distance (small compared to the distance between the atomic sources and the wafer, which is $5 \sim 10$ cm) the thickness gradient, implemented by keeping the substrate stationary, is linear. Hence, the wavelength separation $\Delta\lambda$ between any two lasers is proportional to the distance between the lasers projected onto $\vec{t}$. For a 2D array with its x-axis making an oblique angle θ with $\vec{t}$ as shown in Fig. 4, we obtained

$$\Delta\lambda_x \propto d \cdot \cos\theta \quad \text{and} \quad \Delta\lambda_y \propto d \cdot \sin\theta, \tag{1}$$

where d is the spacing between neighboring lasers, and $\Delta\lambda_x$ and $\Delta\lambda_y$ are the wavelength separation between neighboring lasers in the x and y direction, respectively. Therefore,

$$\Delta\lambda_y = \tan\theta \cdot \Delta\lambda_x = N \cdot \Delta\lambda_x, \qquad N \equiv \tan\theta. \tag{2}$$

Hence, the wavelength of the $(N+1)$th laser on a row is the same as that of the first laser on the next row, provided $\tan\theta$ is close to an integer. Restricting the array to have N columns, a 2D array with no redundant wavelengths is obtained. To obtain a large N, θ must be large and close to 90 °C. The number of rows in such an array, M, can be made very large. Ideally, M is limited only by the total gain bandwidth (~ 100 nm),[16] which has to be larger or equal to the total wavelength span $(N \cdot M - 1) \cdot \Delta\lambda_x$. However, the actual physical dimension may impose a limit if uniform wavelength separation throughout the array is required. The direction of increasing wavelength rasters through the elements of the array. We shall refer to such an array as a rastered multiple wavelength or RMW laser array.

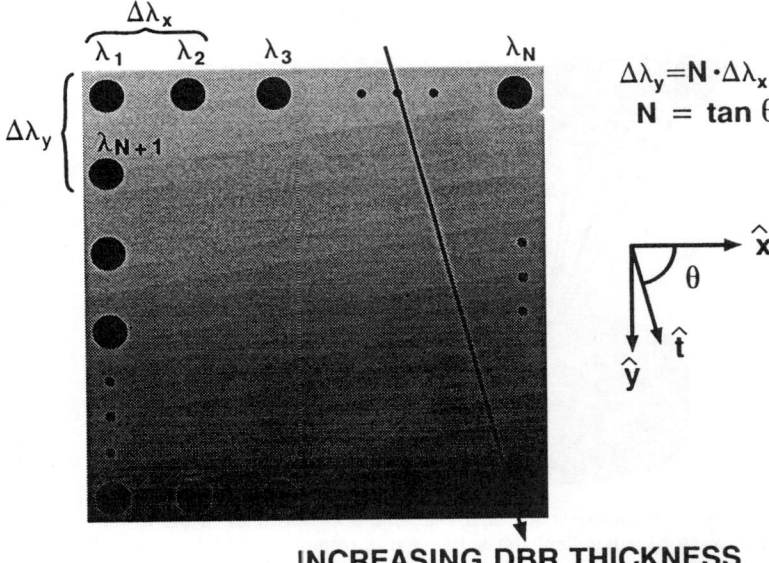

Fig. 4. Schematic of the 2D rastered multiple wavelength (RMW) VCSEL array. By aligning the array obliquely with respect to the direction of thickness variation, a 2D laser array with no redundant wavelengths is obtained.

2.2. Device realization and fabrication

The first step in fabricating a RMW laser array is the MBE growth of the VCSEL structure with spatially chirped layers. The substrate major flat (parallel to the $[01\bar{1}]$ crystal direction) was aligned to the direction intersecting the Ga and Al sources in the MBE chamber. After the MBE growth, the lasers were defined by photolithography. During that process, the photo mask used was aligned so that rows of lasers made an angle θ with the wafer edge which is parallel to the substrate major flat. In our case, the angle θ was chosen to be $\sim 82°$ and $\Delta\lambda_y$ is 7 times $\Delta\lambda_x$. Although larger θ would give a larger N, alignment becomes progressively more difficult.

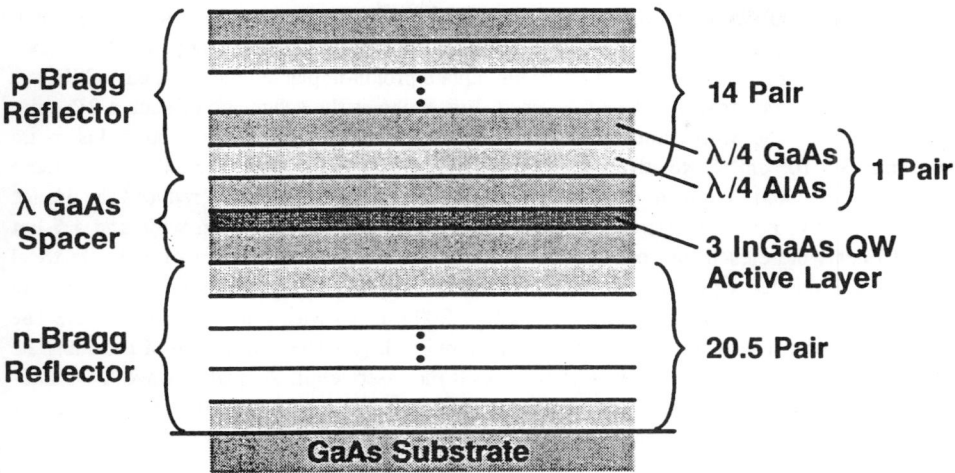

Fig. 5. The schematic of the VCSEL heterostructure used in Secs. 2 and 3.

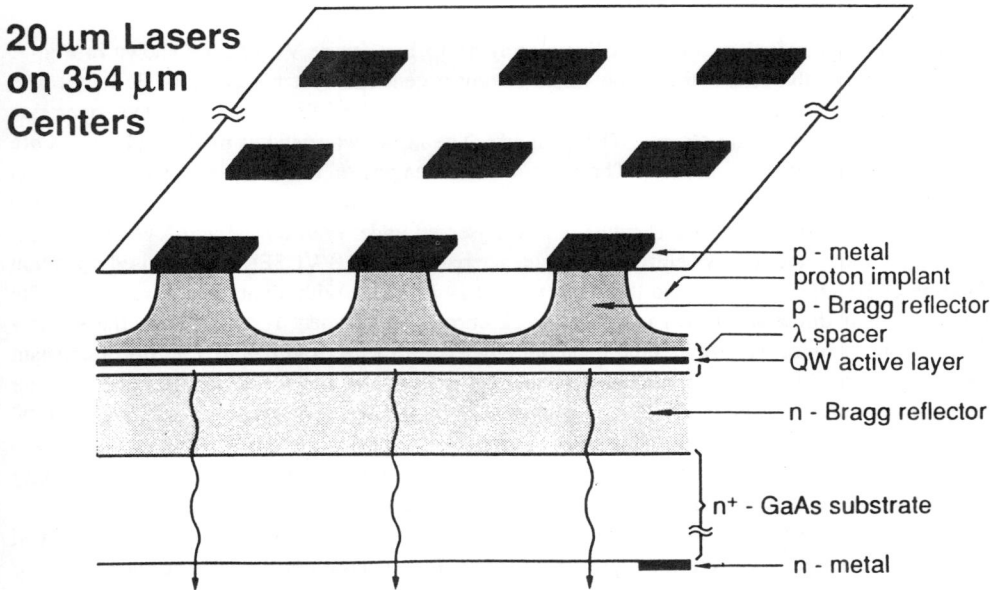

Fig. 6. The schematic of a 2D gain-guided VCSEL array.

The VCSEL heterostructure we used is illustrated in Fig. 5. It is grown on an n^+-GaAs substrate by MBE. The active layer, placed at the center of a one-wave thick $Al_{0.5}Ga_{0.5}As$ layer, consists of three 8 nm $In_{0.2}Ga_{0.8}As$ strained quantum wells. The top and bottom DBRs include 14 and

20.5 pairs of p- and n-doped quarter-wave GaAs/AlAs layers, respectively. 20 nm graded layers were between each GaAs and AlAs layers to provide smoother transition and reduction of series resistance.[9] A phase matching layer consisting of 29 nm AlAs, 20 nm graded layer and 3 nm GaAs cap layer was used on top of the p-DBR mirror to provide proper phase matching with the p-metal contact. High doping density of $4 \times 10^{18}\,\text{cm}^{-3}$ was used for both the p- and n-doped DBRs to further reduce the series resistance. 15 μm or 20 μm square proton-implanted gain-guided VCSEL on 354 μm center-to-center spacings were fabricated. Fig. 6 shows the schematic of such a 2D gain-guided laser array. The proton implant energy and dosage used for the 15 μm square lasers are 100 keV and $3 \times 10^{15}\,\text{cm}^{-3}$. Relatively high threshold currents and voltages are exhibited in these lasers (see next section). We then increased the implant energy to 185 keV and reduced the dosage to $2 \times 10^{15}\,\text{cm}^{-3}$ to make an array of 20 μm square lasers. Moreover, 20 μm wide and 4.5 μm deep grid lines were etched between the lasers using wet chemical etching to provide electrical isolation. Using this latter process, the laser performance improved greatly as will be described in detail in the next section. 20 nm AuBe followed by 200 nm Au was used as the p-contact. The n-metal contact (AuGe/Au) was evaporated everywhere on the polished, back side of the substrate except for 100 μm diameter windows through which the lasers emit. A quarter-wave SiO_x anti-reflection (AR) coating was finally evaporated on the windows to minimize the reflection at the substrate/air interface.

2.3. Experimental measurements of VCSEL array characteristics

The emission spectra of an 8-element linear array of 15 μm square VCSEL are shown in Fig. 7. The inset shows the schematic of the lasers having a center-to-center spacing of 354 μm. The laser spectra are measured at ~ 1.1 times threshold currents under CW (continuous wave) operation with all the lasers emitting a single TEM_{00} mode. The lasing wavelengths of the eight lasers are uniformly separated by ~ 2.6 nm. The emission wavelengths of the lasers under CW operation are approximately 2 nm longer than those under pulsed operation due to heating. However, the same uniform separation of wavelengths is also observed under both pulsed and CW operations.

Fig. 8 shows the light vs current (L–I) characteristics of the 8 VCSEL under pulsed operation (300 ns at 10 kHz). The average threshold current is 14.7 mA with about $\pm 10\%$ variation. The relatively high threshold current is due to weak current confinement resulting from the shallow proton implant used. The differential quantum efficiencies range from 6% to 7%. For a constant drive current of 60 mA, the optical power variation is less than 10%. The variations in the laser threshold currents, quantum efficiencies, and optical power are very small considering the total wavelength span is over 18 nm. This shows that device-to-device uniformity in optical and electrical properties were maintained in spite of the variation in laser cavity thickness. The nonlinearity of the L–I curves is due to the competition of the transverse modes.[6]

A 7×20 RMW array of 20 μm square lasers on 354 μm centers was subsequently fabricated using the procedure described in the previous section. Fig. 9 shows the wavelength distribution of the 2D array. The total wavelength range is as large as 43 nm, from 940 to 983 nm. The average wavelength separation between two neighboring lasers on a row and a column are 0.3 and 2.1 nm, respectively. Each physical position on the 2D array is represented by a unique wavelength. The laser spectra are measured at ~ 1.1 times the threshold currents under CW operation with the lasers emitting a single TEM_{00} mode. All the lasers have a spectral width less than 0.008 nm, the resolution limit of our spectrometer. The wavelength separation can be easily made smaller

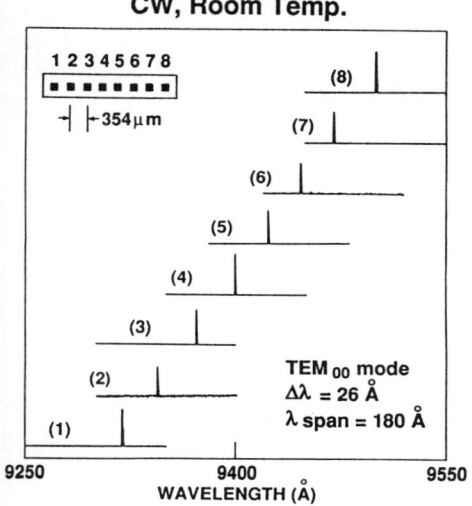

Figure 7. Lasing spectra of an 8-element VCSEL linear array. The inset shows the schematic of the lasers having a center-to-center spacing of 354 μm.

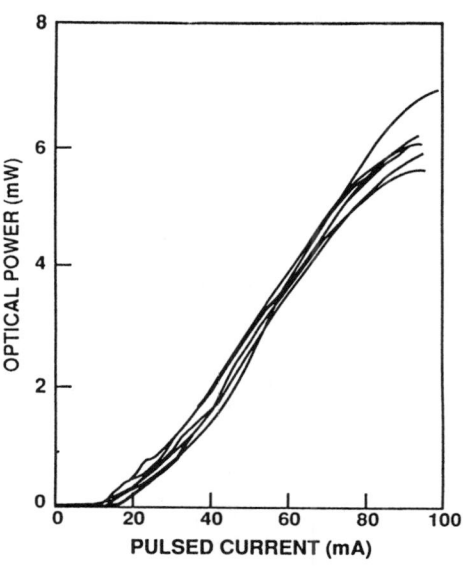

Figure 8. Light vs current characteristics of the eight lasers, whose spectra are shown in Fig. 7.

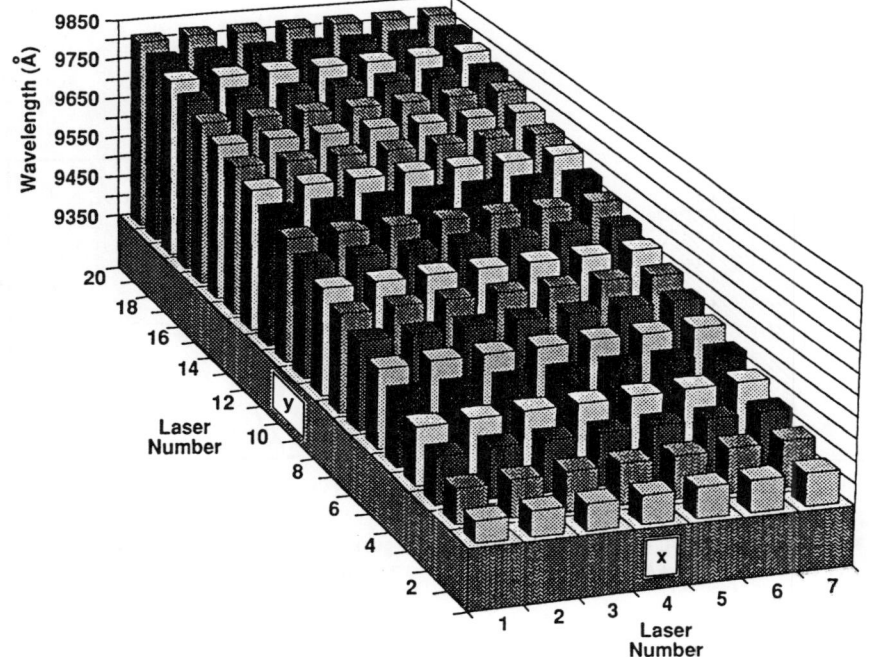

Fig. 9. Wavelength distribution of the 7 × 20 RMW VCSEL array. Each laser emits a unique wavelength. The direction of increasing wavelength rasters through the 2D array.

by reducing the laser spacing or the number of chirped layers. It can also easily be made larger by doing the opposite. Fig. 10 shows the wavelength distribution replotted as a function of laser number K, where $K = (X-1) + 7(Y-1) + 1$, $X = 1, 2, \ldots, 7$, and $Y = 1, 2, \ldots, 20$. A nearly linear relation is obtained indicating uniform wavelength separation and rastering between the lasers.

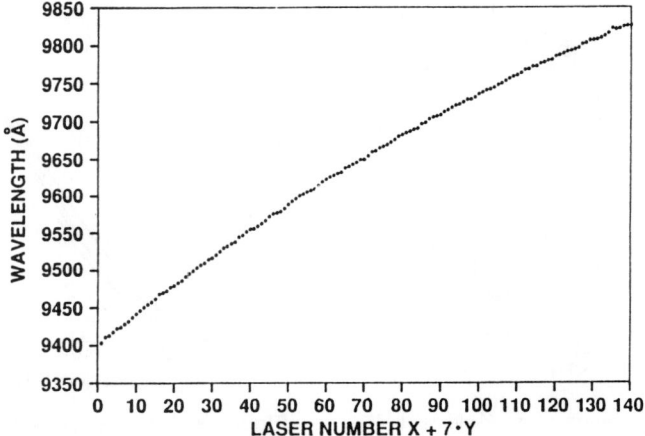

Fig. 10. The laser wavelength as a function of K. A new laser numbering sequence K defined as $(X-1) + 7(Y-1) + 1$, where $X = 1, 2, \ldots, 7$ and $Y = 1, 2, \ldots, 20$.

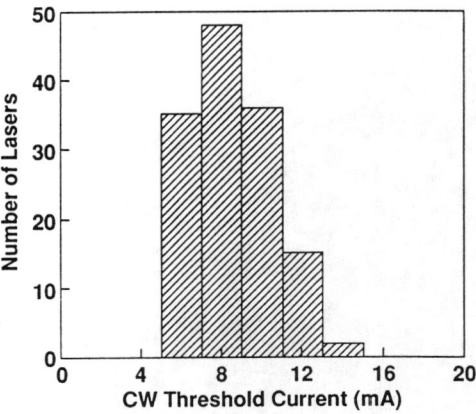

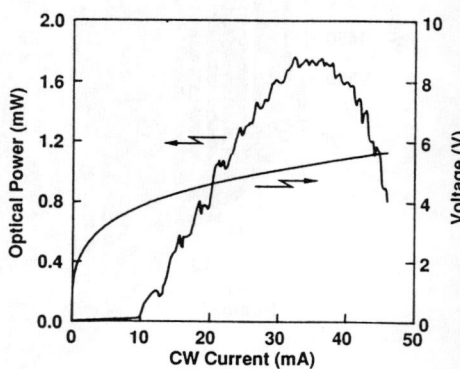

Figure 11. Distribution of CW threshold currents of the 7 × 20 RMW wavelength VCSEL array.

Figure 12. Light intensity vs current and voltage vs current characteristics for a typical VCSEL chosen randomly, $K = 22$, under CW operation.

Fig. 11 shows the CW threshold current distribution of the 2D RMW laser array. The average CW threshold current is 8.5 mA. The 2D laser array exhibits good uniformity with 85% of the lasers having threshold currents within ±3 mA of the average values. At thresholds, the average voltage and the differential series resistance are 3.7 V and 70 Ω, respectively. The threshold currents, voltages, and resistances are all greatly reduced due to deeper proton implant and better isolation between the lasers. The light intensity vs current (L–I) and voltage vs current (V–I) characteristics for a typical laser in the array chosen randomly (K = 22) is shown in Fig. 12. The threshold current and voltage are 9.9 mA and 3.8 V, respectively. The series resistance is about 44 Ω for current between 15 to 45 mA. Using spectrally resolved near field measurements, we identified that the resonances in the L–I curve was due to a combination of residual reflection at the substrate/air interface[11] and transverse mode competition. High CW output power of ~ 1.8 mW is obtained with 35 mA drive current. At higher currents, the optical power is reduced due to heating. However, the laser is not damaged and the L–I curve is repeatable.

2.4. Four-channel WDM system experiment

Fig. 13 shows the WDM system experimental setup. In this experiment, we used a 4 × 1 linear VCSEL array with a center-to-center spacing of 508 μm. These lasers are located on the lines running diagonally with respect to the x and y axis. We chose this relatively large physical spacing between the lasers because it yields a wavelength separation of 1.5 nm, limited by the passband of the optical filter used at the receiving end. The lasers used in this experiment were not packaged on heat-sunk; the substrate was positioned on a sapphire window, through which the output beams were collected. Nontheless, simultaneous operation of four lasers was attained. The lasers were modulated with $2^{15} - 1$ word-length non-return-to-zero (NRZ) pseudo-random patterns at 155 Mb/s and the output light was butt-coupled into four 50/125 multimode fibers (MMF) that were fixed in a silicon V-groove holder. The typical coupling efficiency was 3 dB, limited by the beam divergence through the substrate and a sapphire window (450 μm total thickness). The variation in the coupling efficiency for the four fibers was less than 1.5 dB. Higher coupling efficiency is expected with a more optimized laser design and the use of lensed fibers.[7] The cross-coupling from the adjacent 508 μm-spaced lasers into the optical fiber was less than −60 dB. The signals were combined into a single fiber with a star coupler and transmitted over approximately 20 m of fiber. The maximum output power from each laser was approximately −6 dBm with simultaneous CW operation of four lasers. The lasers were biased at $1.4 I_{th}$, corresponding to ~ 12 mA, under CW operation. A greater output power is expected with proper heatsinking of the lasers.

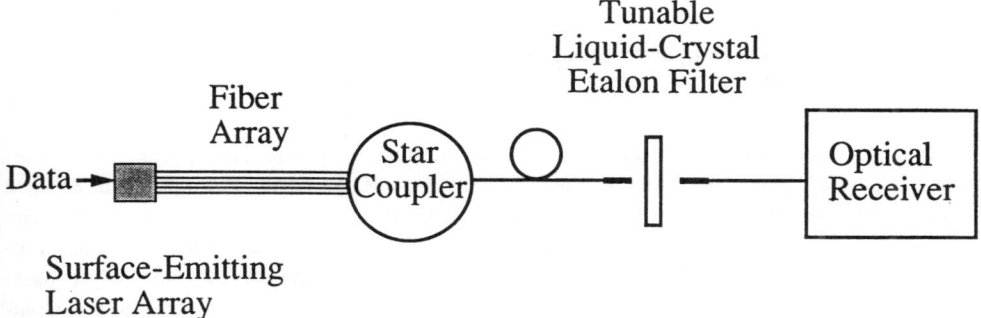

Fig. 13. WDM system experimental setup. The darkened elements indicate the four lasers used in our experiment.

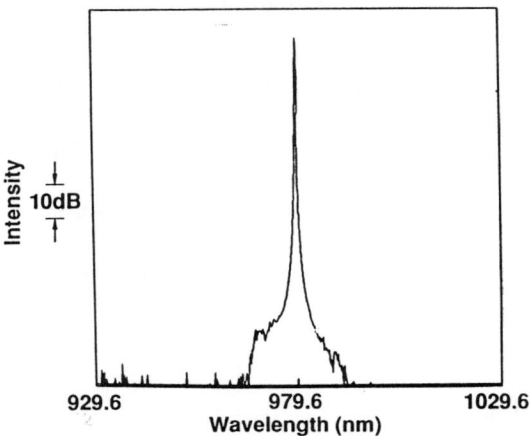

Fig. 14. Optical spectrum of a typical VCSEL.

Fig. 14 shows the optical spectrum of a single laser. The spectral width is resolution-limited as measured by both an optical spectrum analyzer, as shown here, and a spectrometer with a resolution of 0.008 nm. The spectrum revealed single-longitudinal mode behavior due to the extremely short cavity length ($\sim 1~\mu$m) of the lasers. The sidemode suppression was larger than 45 dB, hence the optical crosstalk due to sidemodes was negligible. Furthermore, the optical spectrum broadens very little when the laser was modulated at 155 Mb/s. All of these indicate the VCSEL's are highly suitable for high-density WDM based experiments.

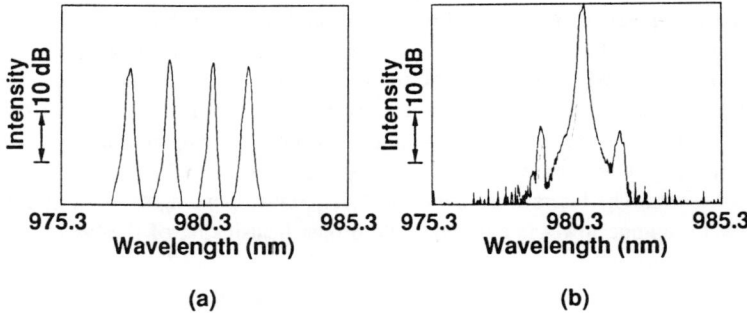

Fig. 15. The optical spectrum of (a) four multiplexed SEL laser outputs, and (b) the liquid-crystal etalon filtered output.

Fig. 15(a) shows the optical spectrum at the output of the star coupler. Four uniformly separated peaks with nearly equal intensities were obtained. The optical filter used in the experiment was an electronically tunable liquid-crystal Fabry–Perot etalon[17] that can also be potentially constructed and operated as a 2-dimensional filter array. The filter output is shown in Fig. 15(b). The filter had a free-spectral-range of 18 nm and a passband of approximately 0.2 nm. The power ratio of the filtered signal to the rejected channels is more than 25 dB. The filtered light was detected by an InGaAs receiver and the bit-error-ratio (BER) was measured to study the crosstalk between the multiplexed signals.

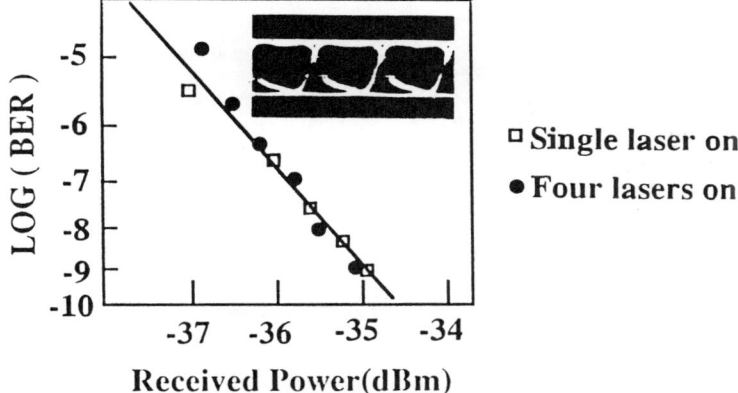

Fig. 16. Bit-error-ratio vs received power for the case when just one laser was operated as compared with when all four lasers were on. Inset shows an open eye diagram at $<10^{-9}$ BER.

Fig. 16 shows the BER as a function of received power for the case when only one laser is modulated as compared with the case when all four lasers are simultaneously operated. The well-overlapped lines for both cases show that no sensitivity penalty was measured with simultaneous operation of four lasers. This indicates that the electrical or optical crosstalk between the lasers is negligible. A BER of 10^{-9} was achieved with a received power of -35 dBm. An open eye diagram was obtained, as shown by the inset of Fig. 16, for lasers modulated with $2^{15} - 1$ word-length NRZ pseudo-random code at 155 Mb/s with less than 10^{-9} BER. The bit rate used in our experiment was limited by the impedance mismatch between the lasers and the drive circuitry and the probes used to drive the lasers.*

2.5. Discussions

2.5.1. Other RMW laser array design approaches

In the foregoing sections we have introduced a new device, namely, a 2D rastered multiple wavelength laser array. We have shown that by implementing spatially chirped Bragg reflectors in a slantingly aligned array, a RMW VCSEL array with uniform wavelength rastering can be achieved. The physical implementation of the chirped layers was done by keeping the wafer stationary during the MBE growth of 2 pairs of DBR layers. This technique offers simplicity and elegance. It is worth noting that many large 1D arrays can be fabricated on the same wafer having the same central wavelength and wavelength separation. As for 2D arrays, many with the same wavelength range can be obtained simply by stacking them as shown in Fig. 17. Nonetheless, the arrays made from the portion closer to Ga and Al sources will be different from the ones made from the portion away from the sources. This can be overcome by using tunable lasers with either discrete tuning[11] or with continuous tuning as will be discussed next.

* Recently, we have demonstrated a monolithic high-speed 16-wavelength laser array with each laser having >5 GHz bandwidth. A record bandwidth of 80 Gb/s was obtained. This result was reported in **Conference on Lasers and Electro-Optics**, Baltimore, MD, May, 1991, pp. 575–576, and will be published soon.

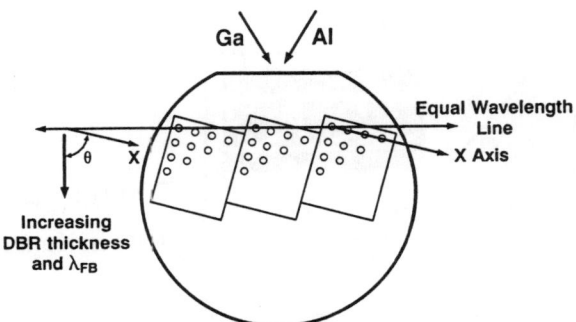

Fig. 17. Schematic of 2D wavelength rastered laser arrays on a whole wafer.

The 2D array with unique wavelengths can also be obtained by variations on our approach. The spacing between the lasers can be varied in both x and y directions independently. As indicated in equation (1), the wavelength separation will be varied accordingly. Alternatively, two dimensional chirped layers can be obtained by keeping the wafer stationary for one layer and then a few more layers with the wafer rotated by 90°.

We also note that the spatially tapered layers can also be made with a number of alternative techniques. For example, by using growth on a nonplanar substrate,[18] VCSEL's with different layer thicknesses can be obtained since the growth rate depends on the size of the pattern, e.g. mesas, on the substrate. Another approach is to pattern etch either the substrate holder (the Mo block)[19] or the back side of the substrate to provide a temperature gradient, which then transfers into a growth gradient on the substrate. Using MOCVD (metal-organic chemical vapor deposition) as the growth technique, laser-induced desorption[13] may be used to produce the thickness gradient also.

2.5.2. New applications enabled by RMW laser arrays

There are a number of applications which are enabled by this new invention. First, very high data transmission throughput can be made available with the additional dimension of freedom of wavelength. Even if each laser diode is operated at relatively slow rate (K Gbit/s), with a large number of wavelengths, N, the total throughput simply multiplies and becomes $N \times K$ Gbit/s. The fact that fairly slow ($\sim$ 1 GHz) diode lasers and electronics can be used to achieve high data rate makes the designs and packaging of lasers, laser drive circuitries, and receivers all much easier and cost-effective. One possible application of such high data rate link is computer-to-computer interconnects. By representing each bit of a byte by one or even two wavelengths, a byte-wide parallel WDM transmission can be obtained.[20–22] This allows very high speed interconnection between large computers, which would be otherwise very costly and difficult to achieve.

The RMW laser arrays provide extra capabilities also for board-to-board free-space optical interconnects. Among the configurations for board-to-board interconnection, the broadcast scheme and imaging interconnection using a holographic element appear to be very promising.[23] For the former, with the use of a RMW array, parallel broadcast can be obtained, which improves the speed tremendously. On the other hand, for the imaging interconnection approach, the wavelength diversity can lead to reduction of crosstalk.[24] With each laser being made wavelength-tunable, the interconnection is thus reconfigurable.

A widely tunable electrical source can also be generated with the use of heterodyning any two lasers with different wavelengths. An electrical beat signal whose frequency equals to the wavelength difference can be obtained. The tuning range of the electrical signal is expected to be very wide, from MHz to THz. Such a tunable electrical source can be very important for a variety of applications in electronic instrumentations.

Using the entire array as an ensemble of optical source, a fast wavelength-tunable laser with a large tuning range is obtained. The tuning speed in this case can be faster than 1 ns, since it is simply the speed of each laser. The wavelength step is the wavelength separation of the lasers, which can be made smaller than 0.1 nm. The total wavelength range is limited by the quantum well gain bandwidth and the DBR mirror bandwidth to about 100 nm. Such a tunable laser when further combined with an optical grating yields a beam-steerable optical source capable of resolving hundreds to a thousand points per line.

3. Wavelength-tunable surface emitting laser

Continuously tunable VCSEL is very desirable as described earlier. Previously, we reported a tunable VCSEL using a 3-mirror coupled cavity[11] with a tuning range of 60 Å. In that case, the tuning was discrete and thus its applications are limited. Another tunable VCSEL structure which was proposed earlier involves a laser with an additional vertically stacked quantum well tuning section.[25] The tuning range in that case, however, is limited by the absorption in the quantum well tuning section to an estimated value of a few Å.[25]

Here we report the first experimental result of continuously tunable VCSEL with a tuning span of 18 Å (540 GHz) including a 4 ~ 6 Å blue shift. The tuning was achieved by pumping the tuning current through the n-doped DBR and n^+-GaAs substrate via an additional electrode placed near the laser junction. The junction can be cooled by this current due to the Peltier effect[26-28] or heated by the resistance of the DBR. Thus, the laser emission can be blue shifted or red shifted depending on the tuning current. The laser power remains nearly constant with only 5 dB power variation for most of the tuning range. Although the effect we used here is thermal-electric, the tuning time is relatively fast due to the small size of the cooler. Using time-resolved spectra measurements, we determined the tuning time to be 300 ~ 400 ns.

3.1. Device design and fabrication

The laser heterostructure is the same as described in 2.2. Air-post index-guided 30 μm square VCSEL's were fabricated using wet chemical etching. The post height of the lasers is about 2.7 μm. The etching was stopped just below the active layer and the first pair of n-doped DBR. Fig. 18 shows the schematic of such a VCSEL. Three metal contacts were fabricated on the laser including a p-metal contact on top of the laser post to form the laser drive electrode, an n-metal contact around the laser post to form the ground contact, and another n-contact on the back of the n^+-substrate as the tuning electrode. Proton implant was used everywhere except the inner 20 μm squares on the laser posts and the 20 μm wide rings around the laser posts. The proton implant provides better current confinement in both the lasers and the tuning junctions. The contacts were unannealed with the contact resistances estimated to be about 15 Ω. The lasers emit at ~ 0.99 μm and the output is taken through the polished transparent GaAs substrate. The lasers were tested without heatsinks.

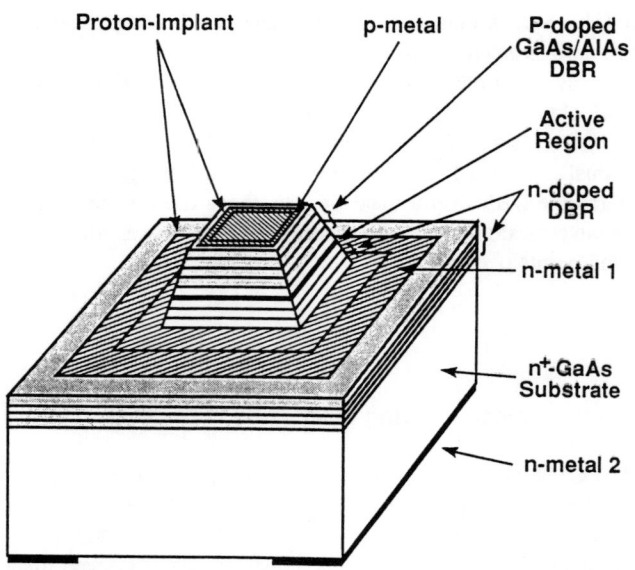

Fig. 18. Schematic of a 2-electrode wavelength-tunable VCSEL.

3.2. Principle of operation

The wavelength tuning mechanism we used here is based on two competing terms. With an electron current flowing from the top n-electrode (n_1) to the bottom n-electrode (n_2), the laser junction heat near n_1 can be piped out towards the n_2 contact. This phenomenon is well known as the Peltier effect or the thermal-electric cooler effect.[26-28] The electron current, on the other hand, also generates heat through the resistance between the two n-electrodes. The total amount of heat W, that would be removed from the laser junction can be expressed as

$$W = \Pi \times I_t - R \times I_t^2 \qquad (3)$$

where Π is the Peltier coefficient of the material, I_t is the tuning current and R is the resistance between the two n-contacts. The direction of positive I_t is defined as from the n_2 to n_1. Hence, for small I_t, the laser wavelength is blue-shifted because of the cooling effect. The amount of blue shift depends on the Peltier coefficient, the resistance and the temperature difference between the two n-electrodes. Note, however, there would not be any blue-shift for reverse biased I_t. With further increasing tuning current, the resistive heating starts to dominate, and the laser wavelength would thus be red-shifted.

3.3. Experimental results

The CW threshold current of a typical laser is 5.6 mA. The light vs current (L–I) characteristics for the laser using the top n-electrode (n_1) and the bottom n-electrode (n_2) shows basically no difference. However, the current vs voltage (I–V) characteristics for the laser using n_1 as the ground contact as compared with n_2 are substantially different, as shown in Fig. 19(a). The circles indicate the voltages at which the laser reach threshold. The threshold voltages are 4.1 V and 5.1 V, respectively. This threshold voltage reduction with the use of n_1 electrode is expected since the pump current can avoid traveling through many potential barriers in the n-doped Bragg reflectors. Indeed, a back-to-back diode type behavior is seen in the I–V characteristics between the two n-contacts, n_1 and n_2, as depicted by Fig. 19(b). The asymmetry in the I–V curve here is attributed to the unequal electrode sizes and the asymmetric graded layers in-between the AlAs and GaAs alternating Bragg reflectors.

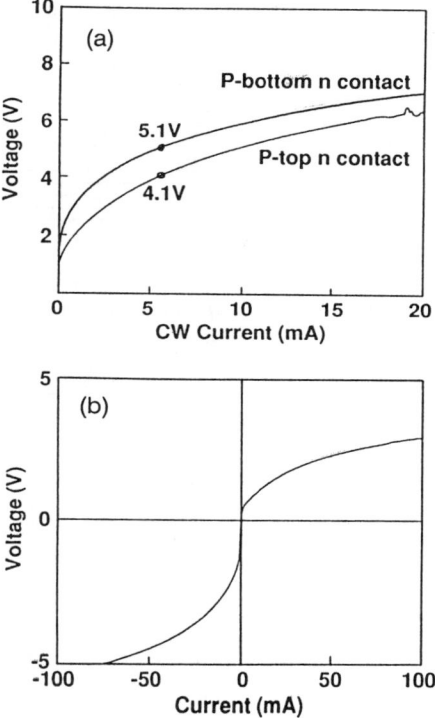

Fig. 19. (a) the current vs voltage (I–V) characteristics for a typical VCSEL using n_1 or n_2 as the ground contact. The circles indicate the voltages at which the VCSEL reaches threshold. (b) typical I–V characteristics between n_1 and n_2.

The laser emission spectra at various tuning current levels with the laser drive current biased at 2.5 I_{th} under CW, room temperature operation is shown in Fig. 20. Trace 1 to 10 were measured with the tuning current being 0, 5, 14, 31, 37, 44, 51, 58, 64, and 67 mA. The laser emits nearly a single mode and no mode hop was observed throughout the tuning range. As described previously, the laser wavelength first blue-shifts due to the Peltier effect with increasing I_t. As I_t further increases, the laser wavelength turns around and red-shifts due to resistive heating between the n-contacts.

Fig. 21 shows the wavelength-shift as a function of the tuning current at two laser drive current levels. The curve qualitatively agree with what is expected from Eq. (3). Since the amount of blue-shift depends on the temperature difference between the laser junction and the back of the

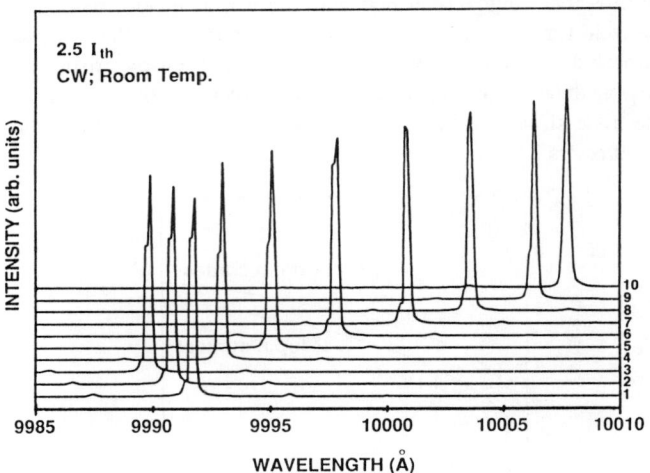

Fig. 20. VCSEL emission spectra at various tuning current levels with the laser drive current biased at 2.5 I_{th} under CW, room temperature operation. Trace 1 to 10 were measured with the tuning current being 0, 5, 14, 31, 37, 44, 51, 58, 64, and 67 mA.

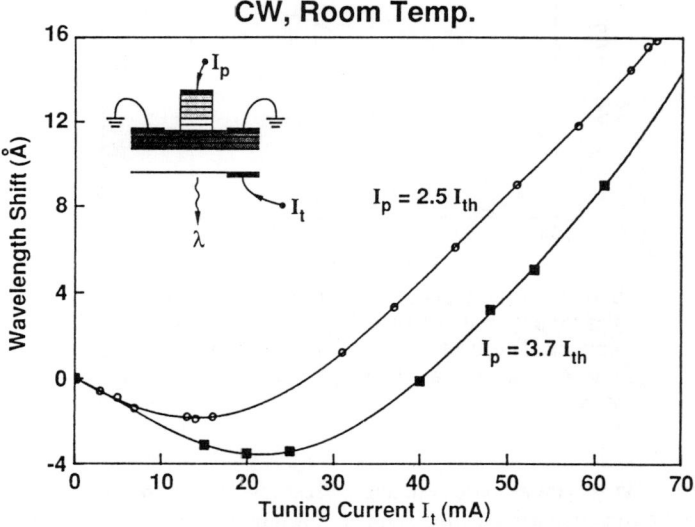

Fig. 21. Wavelength-shift as a function of the tuning current at two laser drive current levels.

substrate, a larger blue-shift for the higher laser drive current was obtained. The amount of blue-shift we typically observed is 4 ~ 6 Å and that of red-shift is about 16 Å. The tuning rate is approximately 0.18 Å/mA (5.4 GHz/mA) and 0.45 Å/mA (13.5 GHz/mA) for the blue-shift and red-shift regime, respectively. The blue-shift is expected to be larger when the back side of the substrate is properly heatsunk. The net blue-shift and the tuning rate we obtained here are both much larger than those of the previously reported tunable edge-emitting lasers using similar effect.[26,27] This is attributed to the use of the Bragg reflectors in between the two n-contacts. For VCSEL's fabricated with larger post-heights such that the top n-contacts are on the GaAs substrate, very negligible amount blue-shift is obtained. We do not understand this observation yet.

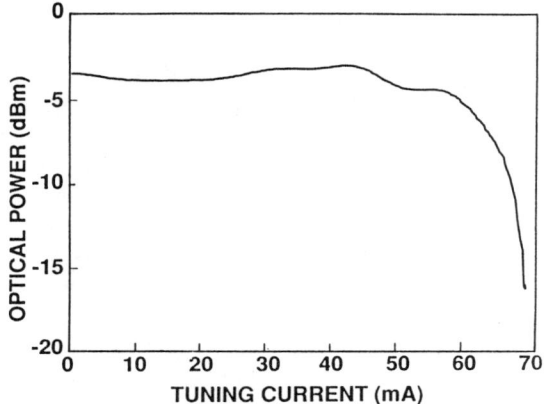

Fig. 22. Typical VCSEL output power as a function of tuning current with the laser biased at a constant current level of 2.5 I_{th} under CW operation.

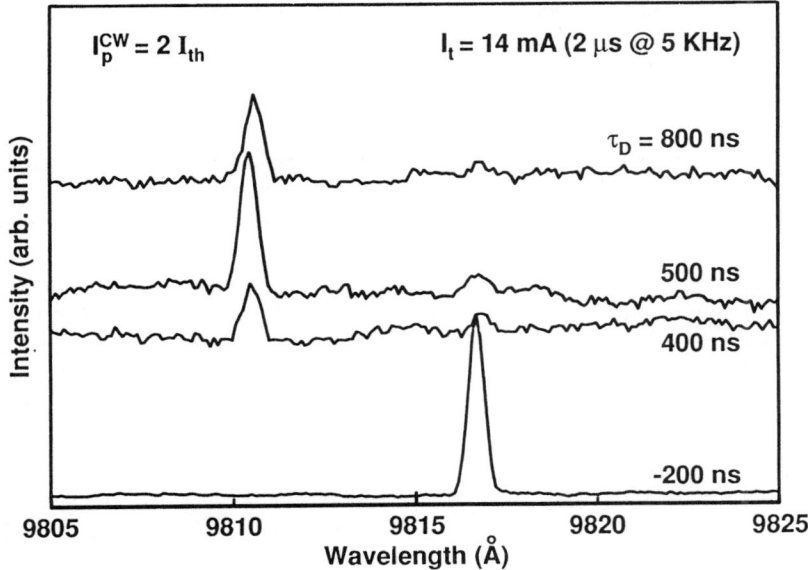

Fig. 23. Time-resolved spectra for a typical VCSEL biased at 2I_{th} under CW operation measured with 100 ns gates at various time delay τ_D.

Fig. 22 shows the laser output power as a function of tuning current with the laser biased at a constant current level of 2.5 I_{th} under CW operation. The output power remains nearly constant throughout most of the tuning range. The output power of this VCSEL with the tuning current turned off is −3 dBm. The optical power is fairly insensitive to the tuning current for $I_t \lesssim 50$ mA. For I_t greater than 60 mA, the optical power declines rapidly. The power variation is less than 5 dB for more than 90% of the tuning range.

The tuning speed is determined by measuring the time-resolved spectra, as shown in Fig. 23. Another typical VCSEL was used in this measurement and it was biased at $2I_{th}$ under CW operation. The tuning electrode was driven by 2 μs long square pulses at 5 kHz repetition rate with a peak current of 14 mA. The optical spectra were measured by a boxcar integrator using 100 ns wide gates positioned at various time delay τ_D with respect to the beginning of the tuning current pulses. The tuning current pulses have ∼ 200 ns spikes at both the leading and trailing edges due to the high n_1-n_2 resistance. The measurement resolution of the tuning speed, being limited by the boxcar gate width and the spikes on the tuning pulses, is estimated to be ∼ 200 ns. At $\tau_D = -200$ ns, the laser spectra shows a single lobe at 9816.7 Å. At $\tau_D = 400$ ns, the laser spectra shows a dominant peak at 9810.7 Å, indicating the tuning has occurred. A 6 Å blue-shift was observed for this VCSEL. The emission stays at this blue-shifted wavelength at $\tau_D = 500$ ns, 800 ns and thereafter until the tuning current pulse is turned off.

4. Conclusions

We have invented a generic technique to obtain both one and two dimensional multiple wavelength laser arrays. By implementing spatially chirped layers in the laser cavities, the wavelengths of the lasers are controllably varied. We further demonstrated that a two dimensional spread of wavelengths can be obtained by obliquely aligning the array axes to the direction of the thickness chirp. We achieved 140 nonredundant, unique, single-mode wavelengths from a 7×20 surface emitting laser array. All the lasers exhibit low threshold currents, low operating voltages and series resistances. These lasers emit in the 980 nm regime. The techniques described here, however, are readily extended to both longer, $1.3 \sim 1.5$ μm, and shorter, $0.6 \sim 0.7$ μm, wavelength regimes. The first wavelength-division-multiplexed system experiment using part of the rastered multiple wavelength laser array is also reported. Four lasers were simultaneously operated at a modulation rate of 155 Mb/s. A bit-error-ratio of 10^{-9} was achieved with negligible optical and electrical crosstalk between the lasers.

In addition, we describe a continuously wavelength-tunable VCSEL with a 18 Å tuning range including $4 \sim 6$ Å blue-shift. The laser output power is nearly constant throughout the tuning range with approximately 5 dB variation. The laser exhibits no mode-hopping. The tuning time is determined with time-resolved spectra measurements to be $300 \sim 400$ ns. The amount of blue-shift and the tuning speed are both record values for devices using Peltier effect. This tuning mechanism can be readily implemented onto the multiple wavelength VCSEL array and thus provide precise wavelength control and reconfigurability.

The multiple wavelength VCSEL array should find numerous applications in optical communications and optical interconnects. Very high data rate throughput is expected using the additional dimension of freedom—wavelength. Indeed, an $N \times M$ VCSEL array can be fabricated to have only M independent wavelengths but with a large number, N, of redundancies for each wavelength or to have hundreds, $N \times M$, of unique wavelengths. The former case increases the tolerance on the array reliability, while the latter enhances the system capabilities. The RMW laser array should also be useful for optical imaging and optical signal processing using the unique one-to-one mapping between wavelengths and physical positions of the lasers. In addition, using the RMW laser array as an ensemble, we obtain a fast wavelength switching source with a large tuning range and a fast beam steering source capable of resolving hundreds to a thousand points.

Acknowledgments

The authors gratefully acknowledge helpful discussions and assistance from Chinlon Lin, G. Hasnain (AT and T Bell Labs), C.E. Zah, J.S. Patel, H.A. Johnson, J.A. Walker (AT and T Bell Labs), A. Von Lehmen, N.G. Stoffel and J.R. Wullert.

References

[1] N.W. Carlson, G.A. Evans, D.P. Bour, and S.K. Liew, 1990 *Appl. Phys. Lett.* **56** 16
[2] Z.L. Liau and J.N. Walpole 1987 *Appl. Phys. Lett.* **50** 528
[3] K. Iga, F. Koyama, and S. Kinoshita 1988 *IEEE J. Quantum Electron.* **24** 1845
[4] J.L. Jewell, A. Scherer, Y.H. Lee, S.L. McCall, J.P. Harbison, L.T. Florez, C.J. Sandroff, R.S. Tucker, and C.A. Burrus 1990 *Integrated Photonics Research, Hilton Head SC* paper WD1
[5] R.S. Geels, S.W. Corzine, J.W. Scott, D.B. Young, and L.A. Coldren 1990 *Photonics Technology Lett.* **I2** 234
[6] C.J. Chang-Hasnain, M. Orenstein, A. Von Lehmen, L.T. Florez, J.P. Harbison, and N.G. Stoffel 1990 *Appl. Phys. Lett.* **57** 218
[7] G. Hasnain, K. Tai, J.D. Wynn, Y.H. Wang, R.J. Fischer, B.E. Weir, and A.Y. Cho 1990 *LEOS '90*, paper PD11
[8] C.J. Chang-Hasnain, J.R. Wullert, J.P. Harbison, L.T. Florez, N.G. Stoffel, and M.W. Maeda 1991 *Appl. Phys. Lett.* **58** 31
[9] K. Tai, L. Yang, Y.H. Wang, J.D. Wynn, and A.Y. Cho 1990 *Appl. Phys. Lett.* **25** 2496
[10] H.J. Yoo, J.R. Hayes, N. Andreadakis, E.G. Paek, G.K. Chang, J.P. Harbison, L.T. Florez, and Y.S. Kwon 1990 *Appl. Phys. Lett.* **56** 1942
[11] C.J. Chang-Hasnain, C.E. Zah, G. Hasnain, J.P. Harbison, L.T. Florez, and N.G. Stoffel 1990 *International Semiconductor Laser Conference, Davos, Switzerland, September*
[12] Y. Tokuda, Y. Abe, T. Matsui, N. Tsukada, and T. Nakayama 1987 *Appl. Phys. Lett.* **51** 1664
[13] T.L. Paoli, D.W. Treat, R.L. Thornton, and R.D. Bringans 1990 *Iternational Semiconductor Laser Conference, Davos, Switzerland* 146
[14] M. Nakao, K. Sato, T. Nishida, T. Tamamura, A. Ozawa, Y. Saito, I. Okada, and H. Yoshihara 1989 *Electron. Lett.* **25** 148
M. Nakao, K. Sato, T. Nishida, and T. Tamamura 1990 *IEEE J. Selected Areas in Communications* **I8** 1178
[15] C.E. Zah, K.W. Cheung, S.G. Menocal, R. Bhat, M.Z. Iqbal, F. Favire, N.C. Andreadakis, P.S.D. Lin, A.S. Gozdz, M.A. Koza, and T.P. Lee 1991 *submitted to Optical Fiber Conference, San Diego*
[16] D. Mehuys, M. Mittelstein, A. Yariv, R. Sarfaty, and J.E. Ungar 1989 *Conference on Lasers and Electro-optics, Baltimore* paper FL4
[17] M.W. Maeda, J.S. Patel, Chinlon Lin, J. Horrorbin, and R. Spicer 1990 *Photon. Tech. Lett.* **2**
[18] E. Kapon, J.P. Harbison, C.P. Yun, and L.T. Florez 1988 *Appl. Phys. Lett.* **52** 607
[19] D.E. Bossi, W.D. Goodhue, M.C. Finn, K. Rauschenbach, J.W. Bales, and R.H. Rediker 1990 *Appl. Phys. Lett.* **56** 420
[20] M.L. Loeb and G.R. Stilwell 1988 *IEEE J. of Lightwave Technology* **6** 1306
[21] M.L. Loeb and G.R. Stilwell 1990 *IEEE J. of Lightwave Technology* **8** 239
[22] J. Sauer *private communication.*
[23] H.H. Arsenault, T. Szoplik, and B. Macukow 1989 *Optical Processing and Computing* Chapter 1, Academic Press
[24] E.G. Paek, J.R. Wullert, A. Von Lehmen, A. Scherer, H.Y. Yoo, R. Martin, J.P. Harbison, N.G. Stoffel, L.T. Florez, C.J. Chang-Hasnain, C.E. Zah, K.W. Cheung, M. Orenstein, J.L. Jewell, and Y.H. Lee 1990 *LEOS Annual Meeting, Boston, November, 1990*
[25] N. Yokouchi, F. Koyama, and K. Iga 1990 *Trans. of IEICE* **73** 1473
[26] N.K. Dutta, T. Cella, R.L. Brown, and T.C. Huo 1985 *Appl. Phys. Lett.* **47** 222
[27] S. Hava, R.G. Hunsperger, and H.B. Sequeria 1984 *IEEE J. Lightwave Technology* **2** 175
[28] W.B. Joyce, and R.W. Dixon 1975 *J. Appl. Phys.* **46** 855

Joint Soviet-American Workshop on the Physics of Semiconductor Lasers   May 20–June 3 1991

Semiconductor microcavity effect on spontaneous emission

D.G. Deppe, C. Lei, T.J. Rogers, and B.G. Streetman

Microelectronics Research Center, Department of Electrical and Computer Engineering, The University of Texas at Austin, Austin, Texas 78712-1084 USA

Abstract. Recent experiments on spontaneous emission from semiconductor microcavities have shown that, similar to atomic system, light emission from a semiconductor can exhibit cavity enhanced/inhibited photon emission. An advantage of the semiconductor material system is the advanced technology available in terms of molecular beam epitaxial crystal growth used for the deposition of high quality AlAs/GaAs distributed Bragg reflectors to form the high-Q Fabry–Perot microcavity. Here we discuss experiments demonstrating the enhanced/inhibited spontaneous emission from semiconductor microcavities which are grown using molecular beam epitaxy, and compare the measurements with calculated spontaneous emission rates of the idealized cavity structures.

1. Introduction

The semiconductor vertical-cavity-emitting laser (VCSEL)[1] has attracted considerable attention recently for several reasons. The demonstration of a single quantum well (QW) VCSEL by Jewell and coworkers[2] showed that the reflectivity of AlAs-GaAs Bragg mirrors can be quite high with the potential of ultra-low threshold currents and greatly simplified device fabrication. There is also a growing appreciation for the unique advantages offered by the vertical-cavity geometry over lateral-cavity devices in terms of optical output characteristics, densely packed 2-dimensional arrays, fiber coupling, device testing, and system integration.

Aside from these device considerations, however, there is a physically more fundamental interest in the semiconductor microcavity because these structures operate in a regime where photon emission from the laser is altered by the microcavity itself, thus offering a degree of control over the spatially anisotropic spontaneous emission, and also control over the spontaneous emission rate into preferred cavity modes. The cavity influence on spontaneous emission is concomitant with control over stimulated emission, so that optical gain is also influenced by the semiconductor microcavity. In recent years there has been considerable effort in studying the effects of cavity electrodynamics on spontaneous emission characteristics of atomic systems.[3] There has been speculation that such effects can play a significant role in semiconductor light emitters[4-9], with a few experimental investigations to date.[10-13] Rogers, et al. recently presented data showing a large influence of an AlAs-GaAs distributed Bragg reflector (DBR) on a closely spaced InGaAs QW, with either enhanced or inhibited emission from the QW, for optical modes normal to the cavity, depending on the specific QW to DBR spacing.[12] Yokoyama, et al. have also presented data on semiconductor microcavity effects on spontaneous emission characteristics from QW's, suggesting nearly a factor of 2 decrease in carrier lifetime due to cavity enhanced spontaneous emission.[13]

As compared to experimental investigations of the effects of cavity enhanced/inhibited spontaneous emission, there has been considerably more theoretical work attempting to model such effects. A purely classical model of a dipole placed in front of a reflector and driven by its own reflected field was found to be in agreement with early experimental data on dye molecules closely spaced to metal reflectors.[14,15] However, a quantum mechanical treatment of both the dipole and the electromagnetic field seems preferable, because of the natural accounting for spontaneous emission. The difficulty in handling the semiconductor microcavity's influence on the electromagnetic field comes from the coupling to the external world, or in accounting for the finite Q of the cavity. A common technique in many calculations is to treat the output coupling as a phenomenological loss term leading to constant field decay in the cavity mode of interest. In this case, however, the DBRs of the microcavity are problematic since some penetration depth must be assumed for the photon upon reflection, complicating the case for optical modes in directions other than the cavity normal. A better method in the case of the semiconductor microcavity is to account for the output directly by calculating the electromagnetic field distribution in all of space for optical modes which interact with a dipole contained in the microcavity. Lang, et al. used this technique to model a one-sided, one dimensional laser cavity with full output coupling.[16] Ujihara has also recently analyzed the one-sided, one-dimensional cavity with output coupling to calculate the cavity effect on spontaneous emission.[17] De Martini, et al. also recently described calculations for a Fabri–Perot microcavity which considered output coupling but did not treat distributed reflectors.[18] Here we compare recent experimental data on InGaAs-GaAs QW's closely spaced to AlAs/GaAs DBRs which exhibit the effects of enhanced/inhibited spontaneous emission, with quantum mechanical calculations of the spontaneous emission rates. For the calculated emission rates we take into full account both output coupling and distributed mirrors.

2. Experiment

The crystals studied in this work were grown using molecular beam epitaxy. A schematic representation of three of the structures is shown in Fig. 1. In these structures a single $In_{0.2}Ga_{0.8}As$-GaAs QW of thickness 125 Å is grown closely spaced to a highly reflecting distributed Bragg reflector (DBR) consisting of 19.5 pairs of AlAs-GaAs alternating layers. The spacing between the InGaAs QW and the first interface of the DBR, designated as "d" in Fig. 1, is varied in each of these three structures to achieve an effective length of $\lambda/4$, $\lambda/2$, and $3\lambda/4$, where λ is the optical emission wavelength of the QW in the GaAs crystal. The upper GaAs confining layer is held fixed in the three crystals at a thickness of 0.57 μm (2λ). A fourth crystal with a structure similar to that shown in Fig. 1, but without the full AlAs-GaAs DBR, has also been studied. In this wafer the DBR is replaced with a single AlAs layer of $\lambda/4$ thickness spaced $\lambda/2$ away from the InGaAs QW. All four wafers contain p and n doping in the layers above and below the QW, respectively, to allow electrical current injection into the QW active region. The structures have been characterized using photoluminescence and reflectivity measurements.

Fig. 2 shows the room temperature photoluminescence measured on three of the wafers. Curve (A) shows the spontaneous emission spectrum for the wafer with a QW to DBR spacing of $d = \lambda/2$, while curve (C) is for the wafer with no DBR, (B) shows the spontaneous spectrum of the wafer with $d = \lambda/4$. The spectrum for the $d = 3\lambda/4$ wafer is not shown, since it is similar to the spectrum measured for $d = \lambda/4$. The luminescence is collected normal to the surface of each wafer, and, as is apparent in Fig. 2, both the spontaneous emission intensity and the spectral characteristics are strongly influenced by the DBR. Comparing curves (A) and (B), one can see that a QW to DBR spacing of $d = \lambda/2$ enhances the spontaneous emission from the InGaAs QW (curve (A)), while a $d = \lambda/4$ spacing strongly suppresses the QW emission (C). The enhancement/inhibition ratio for the integrated intensity from the QW for $d = \lambda/2$ to $d = \lambda/4$ spacing is ~ 76. Also, in comparing the spectral curve (A) with the spectral curve (C), it can be seen that the presence of the AlAs-As DBR greatly enhances and narrows the spontaneous spectrum. The spectral narrowing seen in Fig. 2 (A) results from the cavity effect for the case of the InGaAs QW placed between the AlAs-GaAs DBR (which has an estimated reflectivity of $R \sim 0.99$) and the upper GaAs-air interface (which has a reflectivity of $R \sim 0.3$). Note that the cavity does not simply filter the spontaneous emission from the QW. If filtering were the case, all three curves of Fig. 2 would show the same integrated emission intensity. Instead, Fig. 2 shows that the QW spontaneously radiates only into the allowed cavity mode, with the spontaneous linewidth determined by the quality factor, Q, of the cavity. It is difficult at this time to arrive at a quantitative value for the quantitative change in spontaneous lifetime between the three wafers shown in Fig. 2, and more detailed experiments to measure the lifetimes are required. Nonradiative recombination will likely play a role in determining the total recombination lifetime. It is important to recognize, however, that the large change in integrated intensity shown in Fig. 2 between curves (A) and (C) comes directly from the change in probably for electron-hole pairs to spontaneous radiate in the normal direction to the cavity, and that the total spontaneous lifetime for radiative recombination is not fixed solely by the semiconductor material, but is determined by the electromagnetic modes with which the electron-hole pairs interact.

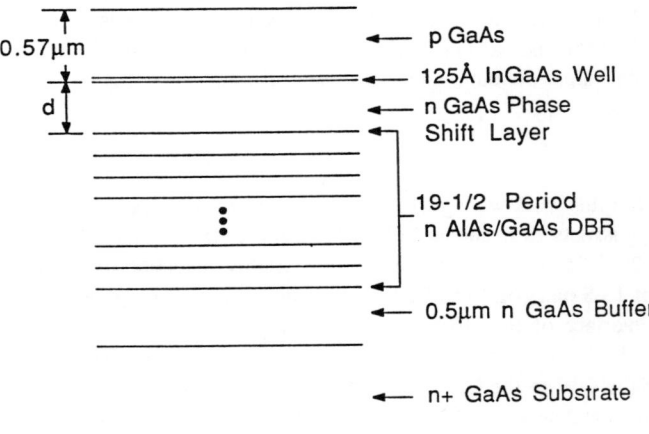

Fig. 1. Schematic representation of the epitaxial crystal structure used in these experiments. The InGaAs QW to AlAs-GaAs Bragg reflector spacing is varied from $\lambda/4$ to $\lambda/2$ to $3\lambda/4$ in three different crystals.

For the epitaxial structure studied here it is of interest to compare the measured linewidth shown in Fig. 2 (A) with the Q expected for the optical cavity composed of the AlAs-GaAs DBR and the GaAs-air interface. The cavity Q is typically given by

$$Q = \nu/\Delta\nu = 2\pi(L_c/\lambda)[-\ln(R_1 R_2)^{1/2}]^{-1} \tag{1}$$

where ν is the cavity resonant frequency, $\Delta\nu$ is the linewidth, L_c is the cavity length, and R_1 and R_2 are the mirror reflectivities. Because the front mirror reflectivity (the GaAs-air interface) is small ($R_2 \sim 0.3$) the cavity Q is not very sensitive to the rear mirror reflectivity value. Therefore, if we take $R_1 = 1$ (for the AlAs-GaAs DBR), $R_2 = 0.3$ and $L_c \sim 2.5\lambda$, we find that Q is ~ 26.

This would correspond to a linewidth $\Delta\nu \sim 1.2 \times 10^{13}$ s^{-1} for our center frequency of $\nu \sim 3.1 \times 10^{14}$ s^{-1}. The actual linewidth found in Fig. 2 (A) is much narrower, however, having a measured value of $\Delta\nu \sim 6.3 \times 10^{12}$ s^{-1}. The discrepancy comes from extended length of the AlAs-GaAs DBR, and to a smaller degree from the spontaneous lineshape independent of the cavity. The Q can also be considered as a measure of the average photon lifetime in the optical cavity for a particular frequency. In this case, the length of the DBR increases the round trip time for the photon in the cavity while still providing a reflectivity of $R_1 \sim 1$. Using Eq. (1), the effective cavity length due to both the GaAs spacer regions and the AlAs-GaAs DBR can be calculated from the measured Q, using Fig. 2 (A), which yields a value of $Q \sim 49$. This measured Q then suggests that the effective cavity length, including the effect of the DBR, is actually $L_c \sim 4.7\lambda$.

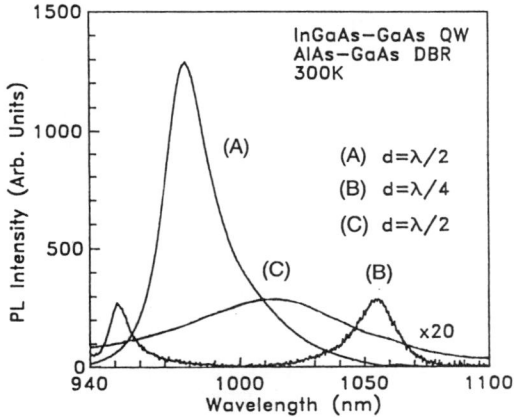

Fig. 2. Photoluminescence spectrum measured normal to the epitaxial surface for InGaAs QW structures with DBR to QW spacing of $\lambda/2$, (A), and $\lambda/4$, (B). (C) shows the photoluminescence from a QW without the DBR.

As stated above, the spontaneous emission from the QW is constrained to radiate into the optical modes of the cavity in which it is placed. Typically, the optical cavities formed in the semiconductor crystal into which a QW may spontaneously radiate is of high loss, and thus of low Q, and of large dimensions with respect to the wavelength of interest. However, as demonstrated here, and in certain vertical-cavity laser structures, this is not always the case. For the short, high-Q optical cavity, or similarly when the QW is placed very close to a highly reflecting mirror, the optical environment in which the spontaneous radiation process takes place is altered. Boundary conditions due to the closely spaced reflectors are imposed upon the vacuum electromagnetic field, the mode density of which controls the spontaneous emission.[19] Since the InGaAs QW has dimensions much less than its optical emission wavelength, it becomes possible to spatially place the QW with respect to the mirror, at either a node or antinode of the vacuum electromagnetic field. Because of the 0° phase shift for a photon reflected from the DBR in the crystal structures studied here, a QW to DBR spacing of $d = \lambda/2$ corresponds to an antinode of the vacuum field.

Similarly, the effect of the DBR on the QW spontaneous emission when the QW is placed at a node of the vacuum field is also pronounced, as can be seen from the spectral curve shown in Fig. 2 (B). In this case the QW to DBR spacing is $d = \lambda/4$, and spontaneous emission from the InGaAs QW is almost completely suppressed in the direction normal to the mirror. In contrast to a gas system, in which the radiating atom samples the entire cavity,[3] placement of the QW at the node of the vacuum modes can force the optical mode density normal to the highly reflecting DBR close to zero.[14] Therefore the cavity inhibition of emission can be greater for the case of fixed placement of the dipole radiator. In Fig. 3 we show a more sensitive measure of the spontaneous spectrum (solid curve), along with the measures reflectivity of the wafer (dashed curve). The detailed influence of the DBR reflectivity curve versus wavelength is observed. The wavelengths at which the DBR has a reflectivity very close to $R \sim 1$ ($\lambda \sim 1$ μm), the spontaneous emission is most strongly suppressed. On the other hand, at the edges of the DBR bandwidth where reflectivity drops, the QW spontaneous emission intensity increases.

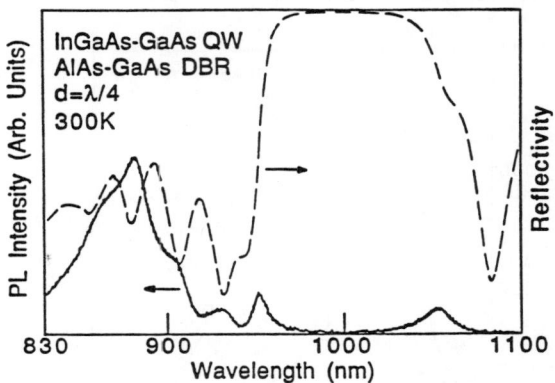

Fig. 3. Both the photoluminescence and reflectivity curves versus wavelength for the QW structure with a QW to DBR spacing of $\lambda/4$. Spontaneous emission is greatly suppressed for wavelength regions where the DBR attains high reflectivity.

3. Calculated spontaneous emission rates from the microcavity

The spontaneous lifetimes, τ_{sp}, of a dipole in the semiconductor microcavity depends on the summation of the emission rates of the dipole into every optical mode with which it interacts, written as,

$$(\tau_{sp})^{-1} = A_{sp} = \sum_{k} A_{k}^{0}, \qquad (2)$$

where A_{sp} is the total spontaneous emission rate, and A_{k}^{0} is the relative spontaneous emission rate into a particular optical mode of wavevector k of the cavity. The total emission rate (stimulated plus spontaneous) into each k mode, A_{k}, will be proportional to a matrix element accounting for the interaction between the dipole and the electromagnetic field, expressed by,

$$A_{k} \propto |\langle \Psi^{f}; n'_{k} | \boldsymbol{E}_{k}(\boldsymbol{R}) \cdot \boldsymbol{r} | \Psi^{i}; n_{k} \rangle|^{2}, \qquad (3)$$

where Ψ^{i} and Ψ^{f} are the initial and final wavefunctions of the dipole, $\boldsymbol{E}_{k}(\boldsymbol{R})$ is the electric field operator, $\boldsymbol{R}$ is the position of the dipole, and $\boldsymbol{r}$ is the dipole operator. The electric field operator is given by,

$$\boldsymbol{E}_{k}(\boldsymbol{R}) = U_{k}(\boldsymbol{R})\boldsymbol{u}[a_{k}e^{i\boldsymbol{k}\cdot\boldsymbol{R}} - a_{k}^{\dagger}e^{-i\boldsymbol{k}\cdot\boldsymbol{R}}], \qquad (4)$$

where $U_k(R)$ is a photon normalization factor which carries the structure of the cavity, u is a polarization vector of the field, a_k and $a_k^\dagger$ are the quantum mechanical annihilation and creation operator for the harmonic oscillator mode k. As is standard for the photon emission, we consider only the creation operator, $a_k^\dagger$, which operates on the photon number eigenstate, $|n_k>$, given an emission rate proportional to $n_k + 1$, describing both stimulated and spontaneous emission. For a dipole in an infinite cavity the structure term $U_k(R)$ is taken as an average field amplitude, and becomes proportional to $[\hbar\omega_k/(2\varepsilon L^3)]^{1/2}$, where ε is the dielectric constant, $\omega_k = c'|k|$, c' is the photon velocity in the medium, and L is a dimension of the cavity taken as going to infinity. In the limit of large L the summation in Eq. (2) becomes,

$$\sum_k \to 2(L/\pi)^3 \int k^2 dk \int \sin\theta d\theta \int d\phi = L^3 \int \omega^2/(\pi^2 c'^3) d\omega, \quad (5)$$

where k is taken in spherical coordinates of k, θ, and ϕ, and two polarization modes are accounted for. For large L the cavity dimensions cancel between the field amplitude and the sum over k, making the spontaneous emission rate independent of cavity dimensions. For a small optical cavity the influence of closely spaced mirrors on the field must be retained in the calculation of the spontaneous emission rate, and this influence, as mentioned above, is carried in $U_k(R)$.

The normalization of $U_k(R)$ must also take into account, however, that the semiconductor microcavity has finite Q, since stationary eigenstates for a photon cannot be defined solely in the microcavity. The leaky cavity in 1-dimension has been considered by Lang, et al.[16] and Ujihara.[17] Here we use a similar analysis to calculate the cavity enhancement and inhibition of the spontaneous emission in all directions, for the semiconductor microcavity with structure in one direction. We consider specifically two cavity structures previously investigated experimentally. To simplify the present analysis we also consider an idealized case of a randomly oriented dipole fixed in the cavity at a precise position, R, by means of a QW, and assume that the spontaneous emission proceeds irreversibly, valid for the low-Q cavities.

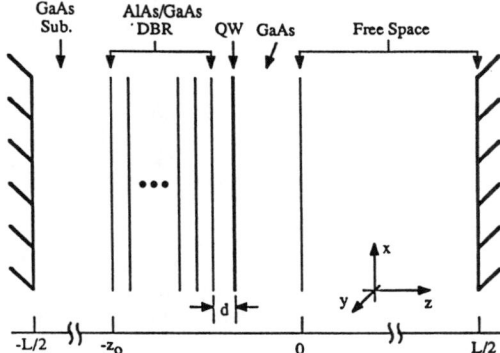

Fig. 4. Schematic showing the cavity structure used in calculating the semiconductor microcavity effect on spontaneous emission.

A schematic of the cavity structure is shown in Fig. 4, in which the semiconductor microcavity is placed in a larger cavity with well defined boundaries at $z = \pm L/2$ (assumed perfect conductors). Two cases are considered by varying the distance, d, between the dipole and the first interface of the AlAs/GaAs Bragg reflector from $\lambda_0/(4\eta)$ to $\lambda_0/(2\eta)$, where $\eta = \sqrt{\varepsilon}$ is the GaAs refractive index, and λ_0 is the free space emission wavelength. The distance from the dipole to the crystal surface in both cases is held fixed at $2\lambda_0/\eta$ to closely resemble the structures investigated experimentally. The larger cavity, extending from $-L/2$ to $+L/2$, takes into account output coupling from the microcavity into free space ($z > 0$) and into the GaAs substrate ($z < -z_0$). The semiconductor microcavity is assumed symmetrical about the z axis, consistent with the epitaxial crystal growth. The normalized structure factor, $U_k(R)$, for each mode is found by solving the Maxwell equations for the electromagnetic field standing waves inside and outside the microcavity in three dimensions, while letting the dimensions of the larger cavity go to infinity in all directions, and keeping the microcavity dimensions fixed in the z-direction, For the present layered structures, Maxwell's equation are most readily solved using matrices.[20] The dielectric constant for the AlAs and GaAs materials are calculated as a function of wavelength using the approximation of Afromowitz.[21] The incremental change with wavevector in the number of modes taken in the z-direction changes from $(\eta L_z/\pi)dk_z'$, where k_z' is in free space, for the case of a large GaAs cavity, to $(L_z/2\pi)(1+\eta)dk_z'$ for the microcavities considered here, reflecting the existence of the non-zero optical mode energy (in the limit of large L_z) in the z-direction in both the GaAs substrate and free space. Some care must be exercised in the normalization of modes which impinge on the GaAs-air surface at greater than or equal to the critical angle, since for these modes the energy of the mode in the region $z > 0$ vanishes relative to that in the GaAs substrate ($z < -z_0$). For these modes the mode density and normalization is determined by the GaAs substrate.

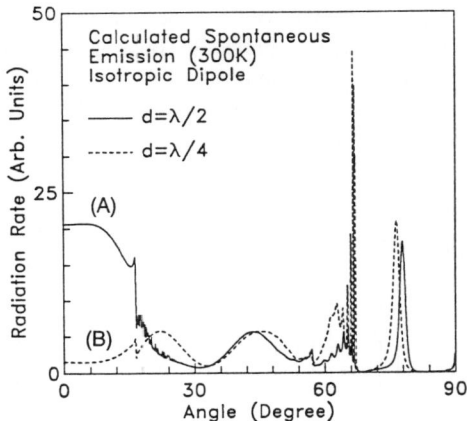

Fig. 5. Calculated spontaneous emission rates into both transverse electric and transverse magnetic modes for the cavity of Fig. 1 with (A) $d = \lambda_0/2\eta$ and (B) $d = \lambda_0/4\eta$.

Fig. 5 shows the calculated integrated emission intensity of the dipole as a function of the angle of emission, relative to the cavity normal, for either the resonant cavity with DBR to dipole spacing of $\lambda_0/2\eta$, Fig. 5 (A), or the antiresonant cavity with DBR to dipole spacing of $\lambda_0/4\eta$, Fig. 5 (B). The emission rates into both transverse electric and transverse magnetic modes are summed for each curve in Fig. 5. The dipole is assumed to emit at a free space wavelength of $\lambda_0 = 1\ \mu$m, with a Gaussian linewidth of 625 Å ($\Delta E \sim 3kT$), approximating the room-temperature photoluminescence measurement conditions of Fig. 2. The amount of cavity enhancement or inhibition in the integrated intensity (integrated over wavelength) which a dipole in the microcavity undergoes is quite sensitive to the linewidth of the spontaneous emission. As considered earlier,[8] this is due to the relative length of the photon coherence as compared to the cavity dimension. Curves similar to those of Fig. 5 but using a spontaneous linewidth corresponding to $T = 77$ K results in much more pronounced peaks and valleys, but with similar qualitative features. We assume for the present calculations of spontaneous emission rates that a summation over many dipoles with closely spaced characteristic wavelengths and narrow linewidths, but producing a broader combined lineshape, is accurately represented by the same number of identical average dipoles which give the same lineshape. This approximation will be most valid for the low level excitations such as used in the experimental investigation ($J_{eq} \sim 10\,\text{A/cm}^2$). The details of band structure and the quasi-Fermi factors, along with the treatment presented here of the microcavity influence on the field, will be required to accurately predict laser operation in similar but higher Q microcavities.[8]

Sharp changes in emission rates with the angle θ are tracked to θ approaching critical angles at the GaAs-air or GaAs-AlAs interfaces, and to shifting of the reflectivity spectrum in wavelength with angle. For both cavities considered, spontaneous emission becomes considerably enhanced beyond the critical angle of the GaAs-air interface because of the increased reflectivity and constructive interference with this interface. We note that because GaAs is used in both the cavity and the substrate there are no waveguide modes, and the greatly reduced emission rates for large angles ($> 65°$) result from poor coupling of radiation modes from the GaAs substrate into the cavity. Integrating the radiation rates of Fig. 5 (A) and (B) gives a relative spontaneous lifetime of a dipole in the two cavities of $[\tau_{sp}(d = \lambda_0/2\eta)]/[\tau_{sp}(\lambda_0/4\eta)] \sim 0.95$ for these relatively low Q cavities.

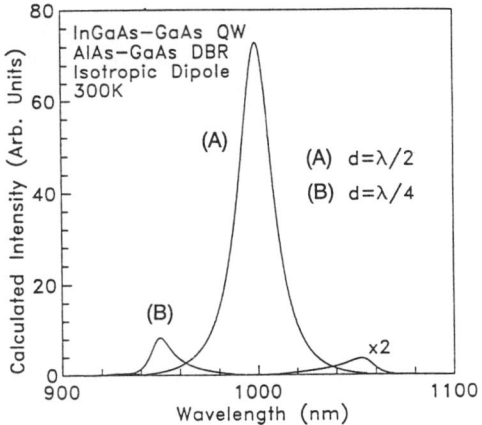

Fig. 6. Calculated spectral characteristics from an InGaAs QW contained in the microcavity of Fig. 5 for (A) $d = \lambda_0/2\eta$ and (B) $d = \lambda_0/4\eta$.

Specific to the calculations reported here is the light collection angle used in the photoluminescence measurement which is $\sim 11°$. This corresponds to a θ (see coordinates of Fig. 4) inside the GaAs crystal of $\sim 3°$, which is to be used in comparing the experimental data of Fig. 2 with the calculations shown in Fig. 5. In this case a relative integrated emission intensity (for $\theta < 3°$) is found by normalizing the total integrated emission rates in each cavity (all angles) by the relative spontaneous lifetimes so that the total emission rates from the two cavities are identical (the same steady state pump rate gives the same steady state loss rate). Under these conditions the relative integrated intensities (from $\theta = 0$ to $3°$) will allow for a comparison with the measured integrated intensities of Fig. 2. The calculated spontaneous spectrum for the two cavities is shown in Fig. 6. It allows direct comparison between calculated spontaneous emission rates and measured. The radiation curves of Fig. 6 predict an increase in the integrated intensity of a factor of 16 for the cavity with $d = \lambda_0/2\eta$ versus the cavity with $d = \lambda_0/4\eta$. This increase can be compared with the photoluminescence measurements of Fig. 2 which give an integrated ratio of 76.

Some difference between predicted and measured intensities may arise due to differences in nonradiative recombination rates. More accurate predictions of spontaneous emission rates from QW's placed in semiconductor microcavities should include the effects of the QW or dipole orientation (operator r in Eq. (2), as well as other details of the electronic band structure and carrier distributions. The position of the dipole in the QW may become important for calculations of low temperature effects when the spread in energy of carrier distribution result in otherwise (large cavity) narrow linewidths. The neglect of QW absorption on cavity Q in the above calculations is valid for angles close to $0°$ because of the low reflectivity of the GaAs-air interface, but may limit spontaneous emission enhancement/inhibition at low drive currents in VCSEL's.

4. Conclusion

The effects of enhanced and inhibited spontaneous emission are found to be important for the semiconductor microcavity through both experimental and theoretical investigations. The control of the spontaneous emission can be quite important for technologically important devices such as the vertical-cavity surface-emitting laser since spontaneous emission represents a parasitic loss in the laser. Additionally, the enhancement of the spontaneous emission into a preferred cavity mode increases the optical gain achievable from the laser active region, making laser operation possible in devices which otherwise may have insufficient cavity Q. It is predicted that the full advantages of the enhanced/inhibited emission effects are yet to be realized, and will come with increased control of cavity optical modes as well as control over the electron-hole dipole.

This work has been supported through The University of Texas research funds (DGD) and the Joint Services Electronics Program contract no. AFROSR F 492620-89-C-044 and the Army Research Office contract no. DAAL 03-88-K-0060 (TJR and BGS).

References

[1] F. Koyama, S. Kinoshita, and K. Iga 1989 *Appl. Phys. Lett.* **55** 221
[2] J.L. Jewell, A. Scherer, S.L. McCall. Y.H. Lee, S. Walker, J.P. Harbison, and L.T. Florez 1989 *Electron. Lett.* **25** 1123
[3] See, for example, S. Haroche and D. Klepper, and references therein 1989 *Phys. Today* **42** N1 24
[4] J.P. Wittke 1975 *RCA Rev.* **36** 655
[5] E. Yoblonovitch 1987 *Phys. Rev. Lett* **58** 2059
[6] H. Yakoyama and S.D. Brorson 1989 *J. Appl. Phys.* **66** 4801
[7] S.D. Brorson, H. Yokoyama, and E.P. Ippen 1990 *IEEE J. Quan. Electron.* QE-26 1492
[8] D.G. Deppe 1990 *Appl. Phys. Lett.* **57** 1721
[9] Y. Yamamoto, S. Mashida, W. Richardson, K. Igeta, and G. Bjork 1990 *Conference on Quantum Electronics, Anaheim, CA* paper QWC6
[10] E. Yoblonovitch, T.J. Glitter, and R. Bhat 1988 *Phys. Rev. Lett* **61** 2546
[11] D.G. Deppe, J.C. Campbell, R. Kuchibotla, T.J. Rogers, and B.G. Streetman 1990 *Electron. Lett.* **26** 1665
[12] T.J. Rogers, D.G. Deppe, and B.G. Streetman 1990 *Appl. Phys. Lett.* **57** 1858

[13] H. Yokoyama, K. Nishi, T. Anan, H. Yamada, S.D. Brorson, and E. Ippen 1990 *Appl. Phys. Lett.* **57** 2814
[14] K.H. Drexhage 1974 *Progress in Optics, edited by E.Wolf (North Holland, Amsterdam) XII* 163–232
[15] R.R. Change, A. Prock, and R. Silbey (Wiley, New York, 1978) Advantages in Chemical Physics, edited by I.Prigogine and S.A.Rice *XXXVII* 1–65
[16] R. Lang, M.O. Scully, and W.E. Lamb 1973 *Phys. Rev. A.* **7** 1788
[17] K. Ujihara 1988 *IEEE J. Quan. Electron.* **QE-24** 1367
[18] F.De Martini, M. Marrocco, P. Mataloni, L. Crescentini, and R. Loudon 1991 *Phys. Rev. A.* **43** 2480
[19] R. Loudon 1983 *Oxford University Press, New York* 5
[20] M. Born and E. Wolf 1980 *Pergamon Press, New York* 1
[21] M.A. Afromowitz 1974 *Solid State Comm.* **15** 59

Frequency stabilized diode lasers

V.L. Velichansky
P.N. Lebedev Physical Institute, Academy of Science of the USSR,
53 Leninsky pr., 117924 Moscow, USSR

Abstract. Frequency stabilized semiconductor lasers have many applications in spectroscopy, metrology and coherent communication systems. The last trends and results on frequency stabilization of laser diodes are presented in the report.

Three laser systems are considered: a solitary laser diode, a diode with integrated composite cavity, and a semiconductor laser with external optical feedback. The number of parameters to be stabilized and the required stability of the parameters are compared for these systems.

The transmission (reflection) resonances of high-Q cavities, the absorption (both Doppler limited and Doppler free) lines of alkali-, alkali-earth metals, and some other atomic and molecular transitions are analyzed as possible references for frequency stabilization of III–V lasers.

Different techniques of frequency (and frequency difference) stabilization are described with the best results achieved.

Authors' index

Alferov Zh.I.	vii	Kish F.A.	1
Andreev V.M.	24	Kochergin A.V.	6
Avrutin E.A.	58	Kochnev I.V.	49
Ayling S.G.	111	Kozlov Yu.G.	75
Bisberg J.	37	Larionov V.R.	24
Bogatov A.P.	85	Lei C.	172
Botez D.	14	Maeda M.W.	151
Bradshaw S.A.	111	Marsh J.H.	111
Bringans R.D	104	Nabiev R.F.	33
Bryce A.C.	111	Neff J.G.	96
Burnham R.D.	1	O'Neil M.	37
Caracci S.J.	1	Paoli T.L.	104
Chang-Hasnain C.J.	151	Pinzone C.J.	96
Chelnokov A.V.	58	Portnoi E.L.	58
Dallesasse J.M.	1	Rafailov E.U.	49
Deppe D.G.	172	Richard T.A.	1
Dupuis R.D	96	Rogers T.J.	172
El-Zein N.	1	Rumyantsev V.D.	24
Eliseev P.G.	33	Shernyakov Yu.M.	49
Epler J.E.	104	Shotov A.P.	87
Filatov I.I.	67	Shubina T.V.	75
Florez L.T.	151	Shvechikov I.Yu.	75
Garbuzov D.Z.	6, 49	Smirnitskii V.B.	37
Gavrilovič P.	37	Singh S.	37
Gorbovitsky B.M.	67	Smith S.C.	1
Gorfinkel V.B.	67	Streetman B.G.	172
Gulakov A.B.	49	Sugg A.R.	1
Gurevich S.A.	67	Sverdlov B.N.	95
Hansen S.I.	111	Timofeev F.N.	130
Harbison J.P.	151	Toropov A.A.	75
Holonyak, N., Jr.	1	Treat D.W.	104
Katsavets N.I.	6	Velichansky V.L.	182
Kazantsev A.B.	24	Yavich B.S.	49
Khalfin V.B.	6, 49	Zory P.S.	86
Khvostikov V.P.	24		

AIP Conference Proceedings

		L.C. Number	ISBN
No. 89	Neutron Scattering – 1981 (Argonne National Laboratory)	82-73094	0-88318-188-6
No. 90	Laser Techniques for Extreme Ultraviolet Spectroscopy (Boulder, CO, 1982)	82-73205	0-88318-189-4
No. 91	Laser Acceleration of Particles (Los Alamos, NM, 1982)	82-73361	0-88318-190-8
No. 92	The State of Particle Accelerators and High Energy Physics (Fermilab, 1981)	82-73861	0-88318-191-6
No. 93	Novel Results in Particle Physics (Vanderbilt, 1982)	82-73954	0-88318-192-4
No. 94	X-Ray and Atomic Inner-Shell Physics – 1982 (International Conference, U. of Oregon)	82-74075	0-88318-193-2
No. 95	High Energy Spin Physics – 1982 (Brookhaven National Laboratory)	83-70154	0-88318-194-0
No. 96	Science Underground (Los Alamos, NM, 1982)	83-70377	0-88318-195-9
No. 97	The Interaction Between Medium Energy Nucleons in Nuclei – 1982 (Indiana University)	83-70649	0-88318-196-7
No. 98	Particles and Fields – 1982 (APS/DPF University of Maryland)	83-70807	0-88318-197-5
No. 99	Neutrino Mass and Gauge Structure of Weak Interactions (Telemark, 1982)	83-71072	0-88318-198-3
No. 100	Excimer Lasers – 1983 (OSA, Lake Tahoe, NV)	83-71437	0-88318-199-1
No. 101	Positron-Electron Pairs in Astrophysics (Goddard Space Flight Center, 1983)	83-71926	0-88318-200-9
No. 102	Intense Medium Energy Sources of Strangeness (UC-Santa Cruz, CA, 1983)	83-72261	0-88318-201-7
No. 103	Quantum Fluids and Solids – 1983 (Sanibel Island, FL)	83-72440	0-88318-202-5
No. 104	Physics, Technology and the Nuclear Arms Race (APS, Baltimore, MD, 1983)	83-72533	0-88318-203-3
No. 105	Physics of High Energy Particle Accelerators (SLAC Summer School, 1982)	83-72986	0-88318-304-8
No. 106	Predictability of Fluid Motions (La Jolla Institute, 1983)	83-73641	0-88318-305-6
No. 107	Physics and Chemistry of Porous Media (Schlumberger-Doll Research, 1983)	83-73640	0-88318-306-4
No. 108	The Time Projection Chamber (TRIUMF, Vancouver, 1983)	83-83445	0-88318-307-2

No. 109	Random Walks and Their Applications in the Physical and Biological Sciences (NBS/La Jolla Institute, 1982)	84-70208	0-88318-308-0
No. 110	Hadron Substructure in Nuclear Physics (Indiana University, 1983)	84-70165	0-88318-309-9
No. 111	Production and Neutralization of Negative Ions and Beams (3rd Int'l Symposium) (Brookhaven, NY, 1983)	84-70379	0-88318-310-2
No. 112	Particles and Fields – 1983 (APS/DPF, Blacksburg, VA)	84-70378	0-88318-311-0
No. 113	Experimental Meson Spectroscopy – 1983 (7th International Conference, Brookhaven, NY)	84-70910	0-88318-312-9
No. 114	Low Energy Tests of Conservation Laws in Particle Physics (Blacksburg, VA, 1983)	84-71157	0-88318-313-7
No. 115	High Energy Transients in Astrophysics (Santa Cruz, CA, 1983)	84-71205	0-88318-314-5
No. 116	Problems in Unification and Supergravity (La Jolla Institute, 1983)	84-71246	0-88318-315-3
No. 117	Polarized Proton Ion Sources (TRIUMF, Vancouver, 1983)	84-71235	0-88318-316-1
No. 118	Free Electron Generation of Extreme Ultraviolet Coherent Radiation (Brookhaven/OSA, 1983)	84-71539	0-88318-317-X
No. 119	Laser Techniques in the Extreme Ultraviolet (OSA, Boulder, CO, 1984)	84-72128	0-88318-318-8
No. 120	Optical Effects in Amorphous Semiconductors (Snowbird, UT, 1984)	84-72419	0-88318-319-6
No. 121	High Energy e^+e^- Interactions (Vanderbilt, 1984)	84-72632	0-88318-320-X
No. 122	The Physics of VLSI (Xerox, Palo Alto, CA, 1984)	84-72729	0-88318-321-8
No. 123	Intersections Between Particle and Nuclear Physics (Steamboat Springs, CO, 1984)	84-72790	0-88318-322-6
No. 124	Neutron-Nucleus Collisions: A Probe of Nuclear Structure (Burr Oak State Park, 1984)	84-73216	0-88318-323-4
No. 125	Capture Gamma-Ray Spectroscopy and Related Topics – 1984 (Int'l Symposium, Knoxville, TN)	84-73303	0-88318-324-2
No. 126	Solar Neutrinos and Neutrino Astronomy (Homestake, 1984)	84-63143	0-88318-325-0
No. 127	Physics of High Energy Particle Accelerators (BNL/SUNY Summer School, 1983)	85-70057	0-88318-326-9
No. 128	Nuclear Physics with Stored, Cooled Beams (McCormick's Creek State Park, IN, 1984)	85-71167	0-88318-327-7
No. 129	Radiofrequency Plasma Heating (Sixth Topical Conference) (Callaway Gardens, GA, 1985)	85-48027	0-88318-328-5
No. 130	Laser Acceleration of Particles (Malibu, CA, 1985)	85-48028	0-88318-329-3

No. 131	Workshop on Polarized ³He Beams and Targets (Princeton, NJ, 1984)	85-48026	0-88318-330-7
No. 132	Hadron Spectroscopy–1985 (International Conference, Univ. of Maryland)	85-72537	0-88318-331-5
No. 133	Hadronic Probes and Nuclear Interactions (Arizona State University, 1985)	85-72638	0-88318-332-3
No. 134	The State of High Energy Physics (BNL/SUNY Summer School, 1983)	85-73170	0-88318-333-1
No. 135	Energy Sources: Conservation and Renewables (APS, Washington, DC, 1985)	85-73019	0-88318-334-X
No. 136	Atomic Theory Workshop on Relativistic and QED Effects in Heavy Atoms (Gaithersburg, MD, 1985)	85-73790	0-88318-335-8
No. 137	Polymer-Flow Interaction (La Jolla Institute, 1985)	85-73915	0-88318-336-6
No. 138	Frontiers in Electronic Materials and Processing (Houston, TX, 1985)	86-70108	0-88318-337-4
No. 139	High-Current, High-Brightness, and High-Duty Factor Ion Injectors (La Jolla Institute, 1985)	86-70245	0-88318-338-2
No. 140	Boron-Rich Solids (Albuquerque, NM, 1985)	86-70246	0-88318-339-0
No. 141	Gamma-Ray Bursts (Stanford, CA, 1984)	86-70761	0-88318-340-4
No. 142	Nuclear Structure at High Spin, Excitation, and Momentum Transfer (Indiana University, 1985)	86-70837	0-88318-341-2
No. 143	Mexican School of Particles and Fields (Oaxtepec, México, 1984)	86-81187	0-88318-342-0
No. 144	Magnetospheric Phenomena in Astrophysics (Los Alamos, NM, 1984)	86-71149	0-88318-343-9
No. 145	Polarized Beams at SSC & Polarized Antiprotons (Ann Arbor, MI & Bodega Bay, CA, 1985)	86-71343	0-88318-344-7
No. 146	Advances in Laser Science–I (Dallas, TX, 1985)	86-71536	0-88318-345-5
No. 147	Short Wavelength Coherent Radiation: Generation and Applications (Monterey, CA, 1986)	86-71674	0-88318-346-3
No. 148	Space Colonization: Technology and The Liberal Arts (Geneva, NY, 1985)	86-71675	0-88318-347-1
No. 149	Physics and Chemistry of Protective Coatings (Universal City, CA, 1985)	86-72019	0-88318-348-X
No. 150	Intersections Between Particle and Nuclear Physics (Lake Louise, Canada, 1986)	86-72018	0-88318-349-8
No. 151	Neural Networks for Computing (Snowbird, UT, 1986)	86-72481	0-88318-351-X
No. 152	Heavy Ion Inertial Fusion (Washington, DC, 1986)	86-73185	0-88318-352-8
No. 153	Physics of Particle Accelerators (SLAC Summer School, 1985) (Fermilab Summer School, 1984)	87-70103	0-88318-353-6

No. 154	Physics and Chemistry of Porous Media—II (Ridge Field, CT, 1986)	83-73640	0-88318-354-4
No. 155	The Galactic Center: Proceedings of the Symposium Honoring C. H. Townes (Berkeley, CA, 1986)	86-73186	0-88318-355-2
No. 156	Advanced Accelerator Concepts (Madison, WI, 1986)	87-70635	0-88318-358-0
No. 157	Stability of Amorphous Silicon Alloy Materials and Devices (Palo Alto, CA, 1987)	87-70990	0-88318-359-9
No. 158	Production and Neutralization of Negative Ions and Beams (Brookhaven, NY, 1986)	87-71695	0-88318-358-7
No. 159	Applications of Radio-Frequency Power to Plasma: Seventh Topical Conference (Kissimmee, FL, 1987)	87-71812	0-88318-359-5
No. 160	Advances in Laser Science–II (Seattle, WA, 1986)	87-71962	0-88318-360-9
No. 161	Electron Scattering in Nuclear and Particle Science: In Commemoration of the 35th Anniversary of the Lyman-Hanson-Scott Experiment (Urbana, IL, 1986)	87-72403	0-88318-361-7
No. 162	Few-Body Systems and Multiparticle Dynamics (Crystal City, VA, 1987)	87-72594	0-88318-362-5
No. 163	Pion–Nucleus Physics: Future Directions and New Facilities at LAMPF (Los Alamos, NM, 1987)	87-72961	0-88318-363-3
No. 164	Nuclei Far from Stability: Fifth International Conference (Rosseau Lake, ON, 1987)	87-73214	0-88318-364-1
No. 165	Thin Film Processing and Characterization of High-Temperature Superconductors (Anaheim, CA, 1987)	87-73420	0-88318-365-X
No. 166	Photovoltaic Safety (Denver, CO, 1988)	88-42854	0-88318-366-8
No. 167	Deposition and Growth: Limits for Microelectronics (Anaheim, CA, 1987)	88-71432	0-88318-367-6
No. 168	Atomic Processes in Plasmas (Santa Fe, NM, 1987)	88-71273	0-88318-368-4
No. 169	Modern Physics in America: A Michelson-Morley Centennial Symposium (Cleveland, OH, 1987)	88-71348	0-88318-369-2
No. 170	Nuclear Spectroscopy of Astrophysical Sources (Washington, DC, 1987)	88-71625	0-88318-370-6
No. 171	Vacuum Design of Advanced and Compact Synchrotron Light Sources (Upton, NY, 1988)	88-71824	0-88318-371-4
No. 172	Advances in Laser Science–III: Proceedings of the International Laser Science Conference (Atlantic City, NJ, 1987)	88-71879	0-88318-372-2
No. 173	Cooperative Networks in Physics Education (Oaxtepec, Mexico, 1987)	88-72091	0-88318-373-0

No. 174	Radio Wave Scattering in the Interstellar Medium (San Diego, CA, 1988)	88-72092	0-88318-374-9
No. 175	Non-neutral Plasma Physics (Washington, DC, 1988)	88-72275	0-88318-375-7
No. 176	Intersections Between Particle and Nuclear Physics (Third International Conference) (Rockport, ME, 1988)	88-62535	0-88318-376-5
No. 177	Linear Accelerator and Beam Optics Codes (La Jolla, CA, 1988)	88-46074	0-88318-377-3
No. 178	Nuclear Arms Technologies in the 1990s (Washington, DC, 1988)	88-83262	0-88318-378-1
No. 179	The Michelson Era in American Science: 1870–1930 (Cleveland, OH, 1987)	88-83369	0-88318-379-X
No. 180	Frontiers in Science: International Symposium (Urbana, IL, 1987)	88-83526	0-88318-380-3
No. 181	Muon-Catalyzed Fusion (Sanibel Island, FL, 1988)	88-83636	0-88318-381-1
No. 182	High T_c Superconducting Thin Films, Devices, and Application (Atlanta, GA, 1988)	88-03947	0-88318-382-X
No. 183	Cosmic Abundances of Matter (Minneapolis, MN, 1988)	89-80147	0-88318-383-8
No. 184	Physics of Particle Accelerators (Ithaca, NY, 1988)	89-83575	0-88318-384-6
No. 185	Glueballs, Hybrids, and Exotic Hadrons (Upton, NY, 1988)	89-83513	0-88318-385-4
No. 186	High-Energy Radiation Background in Space (Sanibel Island, FL, 1987)	89-83833	0-88318-386-2
No. 187	High-Energy Spin Physics (Minneapolis, MN, 1988)	89-83948	0-88318-387-0
No. 188	International Symposium on Electron Beam Ion Sources and their Applications (Upton, NY, 1988)	89-84343	0-88318-388-9
No. 189	Relativistic, Quantum Electrodynamic, and Weak Interaction Effects in Atoms (Santa Barbara, CA, 1988)	89-84431	0-88318-389-7
No. 190	Radio-frequency Power in Plasmas (Irvine, CA, 1989)	89-45805	0-88318-397-8
No. 191	Advances in Laser Science–IV (Atlanta, GA, 1988)	89-85595	0-88318-391-9
No. 192	Vacuum Mechatronics (First International Workshop) (Santa Barbara, CA, 1989)	89-45905	0-88318-394-3
No. 193	Advanced Accelerator Concepts (Lake Arrowhead, CA, 1989)	89-45914	0-88318-393-5
No. 194	Quantum Fluids and Solids—1989 (Gainesville, FL, 1989)	89-81079	0-88318-395-1
No. 195	Dense Z-Pinches (Laguna Beach, CA, 1989)	89-46212	0-88318-396-X
No. 196	Heavy Quark Physics (Ithaca, NY, 1989)	89-81583	0-88318-644-6

No. 197	Drops and Bubbles (Monterey, CA, 1988)	89-46360	0-88318-392-7
No. 198	Astrophysics in Antarctica (Newark, DE, 1989)	89-46421	0-88318-398-6
No. 199	Surface Conditioning of Vacuum Systems (Los Angeles, CA, 1989)	89-82542	0-88318-756-6
No. 200	High T_c Superconducting Thin Films: Processing, Characterization, and Applications (Boston, MA, 1989)	90-80006	0-88318-759-0
No. 201	QED Stucture Functions (Ann Arbor, MI, 1989)	90-80229	0-88318-671-3
No. 202	NASA Workshop on Physics From a Lunar Base (Stanford, CA, 1989)	90-55073	0-88318-646-2
No. 203	Particle Astrophysics: The NASA Cosmic Ray Program for the 1990s and Beyond (Greenbelt, MD, 1989)	90-55077	0-88318-763-9
No. 204	Aspects of Electron–Molecule Scattering and Photoionization (New Haven, CT, 1989)	90-55175	0-88318-764-7
No. 205	The Physics of Electronic and Atomic Collisions (XVI International Conference) (New York, NY, 1989)	90-53183	0-88318-390-0
No. 206	Atomic Processes in Plasmas (Gaithersburg, MD, 1989)	90-55265	0-88318-769-8
No. 207	Astrophysics from the Moon (Annapolis, MD, 1990)	90-55582	0-88318-770-1
No. 208	Current Topics in Shock Waves (Bethlehem, PA, 1989)	90-55617	0-88318-776-0
No. 209	Computing for High Luminosity and High Intensity Facilities (Santa Fe, NM, 1990)	90-55634	0-88318-786-8
No. 210	Production and Neutralization of Negative Ions and Beams (Brookhaven, NY, 1990)	90-55316	0-88318-786-8
No. 211	High-Energy Astrophysics in the 21st Century (Taos, NM, 1989)	90-55644	0-88318-803-1
No. 212	Accelerator Instrumentation (Brookhaven, NY, 1989)	90-55838	0-88318-645-4
No. 213	Frontiers in Condensed Matter Theory (New York, NY, 1989)	90-6421	0-88318-771-X 0-88318-772-8 (pbk.)
No. 214	Beam Dynamics Issues of High-Luminosity Asymmetric Collider Rings (Berkeley, CA, 1990)	90-55857	0-88318-767-1
No. 215	X-Ray and Inner-Shell Processes (Knoxville, TN, 1990)	90-84700	0-88318-790-6
No. 216	Spectral Line Shapes, Vol. 6 (Austin, TX, 1990)	90-06278	0-88318-791-4
No. 217	Space Nuclear Power Systems (Albuquerque, NM, 1991)	90-56220	0-88318-838-4
No. 218	Positron Beams for Solids and Surfaces (London, Canada, 1990)	90-56407	0-88318-842-2

No. 219	Superconductivity and Its Applications (Buffalo, NY, 1990)	91-55020	0-88318-835-X
No. 220	High Energy Gamma-Ray Astronomy (Ann Arbor, MI, 1990)	91-70876	0-88318-812-0
No. 221	Particle Production Near Threshold (Nashville, IN, 1990)	91-55134	0-88318-829-5
No. 222	After the First Three Minutes (College Park, MD, 1990)	91-55214	0-88318-828-7
No. 223	Polarized Collider Workshop (University Park, PA, 1990)	91-71303	0-88318-826-0
No. 224	LAMPF Workshop on (π, K) Physics (Los Alamos, NM, 1990)	91-71304	0-88318-825-2
No. 225	Half Collision Resonance Phenomena in Molecules (Caracus, Venezuela, 1990)	91-55210	0-88318-840-6
No. 226	The Living Cell in Four Dimensions (Gif sur Yvette, France, 1990)	91-55209	0-88318-794-9
No. 227	Advanced Processing and Characterization Technologies (Clearwater, FL, 1991)	91-55194	0-88318-910-0
No. 228	Anomalous Nuclear Effects in Deuterium/Solid Systems (Provo, UT, 1990)	91-55245	0-88318-833-3
No. 229	Accelerator Instrumentation (Batavia, IL, 1990)	91-55347	0-88318-832-1
No. 230	Nonlinear Dynamics and Particle Acceleration (Tsukuba, Japan, 1990)	91-55348	0-88318-824-4
No. 231	Boron-Rich Solids (Albuquerque, NM, 1990)	91-53024	0-88318-793-4
No. 232	Gamma-Ray Line Astrophysics (Paris–Saclay, France, 1990)	91-55492	0-88318-875-9
No. 233	Atomic Physics 12 (Ann Arbor, MI, 1990)	91-55595	088318-811-2
No. 234	Amorphous Silicon Materials and Solar Cells (Denver, CO, 1991)	91-55575	088318-831-7
No. 235	Physics and Chemistry of MCT and Novel IR Detector Materials (San Francisco, CA, 1990)	91-55493	0-88318-931-3
No. 236	Vacuum Design of Synchrotron Light Sources (Argonne, IL, 1990)	91-55527	0-88318-873-2
No. 237	Kent M. Terwilliger Memorial Symposium (Ann Arbor, MI, 1989)	91-55576	0-88318-788-4
No. 238	Capture Gamma-Ray Spectroscopy (Pacific Grove, CA, 1990)	91-57923	0-88318-830-9
No. 239	Advances in Biomolecular Simulations (Obernai, France, 1991)	91-58106	0-88318-940-2